Protecting Groups in Organic Synthesis

Postgraduate Chemistry Series

A series designed to provide a broad understanding of selected growth areas of chemistry at postgraduate student and research level. Volumes concentrate on material in advance of a normal undergraduate text, although the relevant background to a subject is included. Key discoveries and trends in current research are highlighted, and volumes are extensively referenced and cross-referenced. Detailed and effective indexes are an important feature of the series. In some universities, the series will also serve as a valuable reference for final year honour students.

Titles in the Series:

Catalysis in Asymmetric Synthesis
Jonathan M.J. Williams

Protecting Groups in Organic Synthesis
James R. Hanson

Organic Synthesis with Carbohydrates
Geert-Jan Boons and Karl J. Hale

Protecting Groups in Organic Synthesis

JAMES R. HANSON
Reader in Chemistry
School of Chemistry, Physics and Environmental Science
University of Sussex

Sheffield
Academic Press

Blackwell
Science

EXETER COLLEGE
OXFORD

First published 1999
Copyright © 1999 Sheffield Academic Press

Published by
Sheffield Academic Press Ltd
Mansion House, 19 Kingfield Road
Sheffield S11 9AS, England

ISBN 1-85075-957-X

Published in the U.S.A. and Canada (only) by
Blackwell Science, Inc.
Commerce Place
350 Main Street
Malden, MA 02148-5018, U.S.A.
Orders from the U.S.A. and Canada (only) to Blackwell Science, Inc.

U.S.A. and Canada only:
ISBN 0-6320-4506-X

NOTICE: The authors of this volume have taken care that the information contained herein is accurate and compatible with the standards generally accepted at the time of publication. Nevertheless, it is difficult to ensure that all the information given is entirely acccurate for all circumstances. The publisher and authors do not guarantee the contents of this book and disclaim liability, loss, or damage incurred as a consequence, directly or indirectly, of the use and application of any of the contents of this volume.

Trademark Notice: Product or corporate names may be trademarks or registered trademarks, and are used only for identification and explanation, without intent to infringe.

Printed on acid-free paper in Great Britain by
Bookcraft Ltd, Midsomer Norton, Bath

British Library Cataloguing-in-Publication Data:
A catalogue record for this book is available from the British Library

Library of Congress Cataloging-in-Publication Data:
A catalog record for this book is available from the Library of Congress

Preface

The protection of one functional group by masking its reactivity in order to transform selectively another part of a molecule is often the key to a successful organic synthesis. Examination of most syntheses of polyfunctional molecules will reveal the importance of protecting groups and the extent to which they have been used in recent years. Underlying the choice of a suitable protecting group for a particular task must be a sound understanding of the function of the individual protecting group, its stability and the chemistry of both the protection and deprotection steps.

There are encyclopaedic books available listing protecting agents and reviews covering particular groups of compounds. Details of these are provided in the bibliography. It is the object of this book to provide an introduction to the topic, with a mechanistic rationale for the rôle and scope of some of the major protecting groups, so that the postgraduate student or the researcher in the academic or industrial laboratory may appreciate the way to select a specific reagent for a task.

I wish to acknowledge the considerable help given by Professor James Coxon of the University of Canterbury, New Zealand, during the preparation of the manuscript. I would also like to thank Dr Michael Morris of the University of Sheffield for his assistance in the drawing of the chemical structures.

J.R. Hanson
University of Sussex

Contents

1   Introduction

Protecting groups have an essential role in synthesis in masking the reactivity of one functional group whilst allowing the chemical modification of another group in the same molecule. As synthetic targets have become more complex so the need for protecting groups has become greater. The major synthetic problems of 40 to 50 years ago such as that of the steroid hormones involved the construction of ring systems of the correct stereochemistry but the targets had relatively few functional groups. Many of today's targets, such as the polyether and macrolide antibiotics, the terpenoids, such as taxol and forskolin, the nucleosides and peptides, are polyfunctional molecules often with a limited stability. Hence there is a need to introduce control elements into a synthesis to enhance the regioselectivity of particular steps. Protecting groups are important features in these synthetic strategies.

It is possible to establish a series of criteria for a good protecting group.[1-3] Thus, first, the protecting group should be easily and quantitatively introduced under mild reaction conditions that do not disturb other functional groups. Second, the protecting group should be specific for a particular functional group and the protected product should be readily purified. Third, the protected product should be stable to the appropriate reaction conditions and to the purification procedures following the reaction. Fourth, the protecting group should be easily and quantitatively removed under mild conditions which again do not disturb other functional groups. Finally, the deprotected product should be easily separated from the residue of the protecting agent. Whilst such criteria may seem obvious, they are nevertheless, despite many claims to the contrary, rarely completely fulfilled. In this context it is worth remembering that every target structure is unique with its own distinct stereochemical properties and opportunities for neighbouring group participation which may modify the reactivity of a particular centre. Thus what may be a good protecting group in one situation may be useless in another. A computer program has been written to assist in the process of selecting a protecting group for a functional group in a particular sequence.[4]

In selecting a protecting agent, it is important to consider the role of the protecting group in the reaction sequence. For example in protecting an alcohol (**1**), an acetate ester (**2**) may reduce the sensitivity of the alcohol to acid or prevent its oxidation via a chromate ester, whereas a methyl ether (**3**) may prevent the formation of an alkoxide under basic conditions. The nucleophilicity of an amine (**4**) is diminished in its conversion to an amide (**5**), a feature of major importance in the application of amine protecting groups in peptide synthesis.

In many cases a protecting group is required in order to differentiate between two centres of similar reactivity, for example between two hydroxyl groups. In this case steric features may be used to determine which protecting group is used. For example the bulky triphenylmethyl (**6**) or tertiarybutyldimethylsilyl (**7**) groups form ethers more readily with primary alcohols often leaving more hindered secondary alcohols exposed for other reactions.

Many developments[5] in novel protecting groups have been concerned with improvements to the ease with which they can be introduced, to their selectivity and stability and to the facility with which they can be removed. For example a number of methods of activating a carboxyl group in order to facilitate the initial addition of a nucleophile have improved the ease of esterification. Specific reagents such as dicyclohexylcarbodiimide (**8**) have been introduced for this purpose. An alcohol can be activated by its temporary conversion to a better leaving group as in the Mitsunobu reaction. Selectivity in the ease of hydrolysis of esters may

(8)

be introduced by increasing the size of the protecting group. For example
a bulky trimethylacetate (pivaloate) (9) ester is hydrolysed more slowly
than an acetate ester (10). Neighbouring group participation by a hydroxyl group may facilitate the hydrolysis of one ester, allowing another
to remain in place.

(9)   (10)

It is often possible to protect two adjacent centres by involving them in
a cyclic system, for example by the formation of an acetonide (12) from a
vicinal diol (11). A third centre may then be left exposed for reaction.

(11)   (12)

Intramolecular protection by cyclisation through lactone or acetal formation may provide a useful alternative to the addition of a protecting
group.

The manipulation of functional groups by using protecting groups of
contrasting stability is a common strategy.[6] The selection of a protecting
group is often determined by the stability and cleavage of the protecting
group rather than by its introduction.

The concept of orthogonality in protecting groups was developed
by Barany and Merrifield[7] in the course of their work on peptide synthesis. However, it has a much wider application. The general aim of an
orthogonal protecting scheme in a polyfunctional molecule is to be
able to use deprotecting reagents in which the selectivity between the
protecting groups is based on major differences in their chemistry. It
should be possible to cleave members of one orthogonal class of protecting groups in the presence of another class. Several contrasting

sets of protecting groups have been developed.[2, 3, 7] Orthogonal protecting schemes have a particularly important role to play in peptide synthesis where there is a need to protect side-chain functionality as the main peptide chain itself is developed. In the protection of alcohols it is possible to distinguish between protecting groups such as esters (2) that are base-labile and ethers (3) or acetals that are acid-labile.

An understanding of the anomeric effect arising from the interaction of oxygen lone pairs with other bonds has led to a better understanding of the stability of a number of acetals. As a result a series of cyclic ether protecting groups have been developed for use with polyhydroxylic molecules.

Several facets of the chemistry of silicon have played an important role in the development of protecting groups.[8–10] These include the ease with which oxygen–silicon bonds are formed and the ability of the fluoride ion to cleave oxygen–silicon and carbon–silicon bonds selectively (13–14). The steric requirements of different alkyl and aryl groups attached to

silicon have afforded a graded range of protecting groups with a stability ranging from robust to sensitive and a selectivity that can be used to differentiate between sterically hindered and accessible centres.

Carbon–sulfur bonds and benzylic bonds are susceptible to hydrogenolysis and this introduces a further opportunity for differentiating between protecting groups and for removing them under conditions which do not disturb other groups. These methods of deprotection (15–16) can lead to relatively nonpolar, even volatile, residues (17) from the protecting group.

Application of the principles of hard and soft acids and bases and in particular the combination of a hard acid:soft base in which the hard acid bonds to the oxygen atom of an ether or an ester and a soft nucleophile attacks an alkyl carbon (**18**–**19**), has provided a number of useful methods for cleaving protecting groups.[11]

(18)   (19)

Changes in the lability of the protecting group can be brought about by oxidation or reduction. For example by altering a substituent on an aromatic ring by converting an electron-withdrawing nitro group to an electron-donating amino-group one can alter the lability of an aromatic protecting group and thus facilitate its removal. The modification of a protecting group prior to its removal is sometimes known as relay deprotection. Examples of this strategy include the removal of substituted benzyl ethers in the deprotection of alcohols.

Photochemical deprotection[12] is another strategy. This makes use of the different reactivity of the excited state compared with the ground state of a molecule.

Neighbouring group participation, in which an unsaturated system or an adjacent oxygen atom stabilises a carbocation, may also enhance the ease of deprotection. However, neighbouring group participation may also be the source of a problem with a protecting group in that the group, typically an ester, may migrate from one centre to another. There are a number of examples of this in steroid chemistry.[13] Indeed a protecting group can participate in reactions and consequently it cannot always be regarded as an unreactive group. For example, the lone pair on the oxygen atom of the carbonyl of an acetate ester may participate in trans-annular dioxolonium ion formation (**20**–**21**).

(20)   (21)

Fragmentation reactions involving a β-elimination are also used in deprotection. These reactions may be initiated reductively (22–23) or through the use of a trimethylsilyl group (24–25). They can provide a selective deprotection method which allows one protecting group to be differentiated from another. In many instances the fragments of the protecting group are volatile and may be easily removed from the system.

(22)   (23)

(24)   (25)

The facility with which an allyl group coordinates to a metal such as palladium to form a π-allyl complex may also provide a driving force for the selective removal of a protecting group.

Enzymatic methods of deprotection are also attracting interest. They offer the opportunity for chiral discrimination and the resolution of substrates. There are some particularly nice examples of the desymmetrisation of diols by the enzymatic hydrolysis of acetates (26–27).

(26)   (27)

$$R = \text{---- OAc,} \quad \text{Cl}$$

Application of these various principles will be found in the following chapters, which are devoted to protecting groups for individual functional groups.

Many protecting groups form quite cumbersome additions to structures. Consequently, the use of abbreviations such as TBDMS (for tertiarybutyldimethylsilyl) and Cbz (for carbobenzyloxy) has become widespread. The more common abbreviations are listed in the appendix. However, it is worth remembering when using these abbreviations that they can obscure the chemical nature of the functional groups present in the protecting group. Consequently, the use of these abbreviations can obscure the possibility of conflicting reactions involving the protecting group.

2   The protection of alcohols

2.1   Introduction

The need to protect alcohols in polyfunctional molecules has led to the
development of a wide range of protecting groups. In order to select a
protecting group for an alcohol, we need to consider the factors which
underlie the reactivity of alcohols. First, many of their reactions involve
the initial formation of a reactive ester as in substitution reactions to form
an alkyl halide with phosphorus pentachloride or thionyl chloride or in
oxidation reactions with chromium trioxide. Second, on the one hand, a
strong base will lead to the formation of a nucleophilic oxygen anion. On
the other hand, interaction of the lone pairs on the oxygen atom with
mineral or Lewis acids can lead to carbocation formation which may be
followed by the formation of an alkene or by rearrangement. Thus there
are situations in which the hydrogen atom and the lone pairs have to be
protected. Furthermore, many biologically active molecules that are
current synthetic targets possess several hydroxyl groups. It is therefore
often necessary not just to protect a hydroxyl group but also to differentiate between hydroxyl groups by protecting one before manipulating
another.
  Alcohols have been protected as their ethers, acetals and esters, each
derivative possessing its particular advantages and disadvantages. These
will be considered separately in the following sections.

2.2   The protection of aliphatic alcohols as ethers

2.2.1   Methyl ethers

Simple alkyl ethers, such as methyl ethers, require fairly vigorous
conditions for their preparation (CH_3I, Ag_2O; $(CH_3)_2SO_4$, NaOH). Their
synthesis (**1–3**) is based on the formation of an alkoxide (**2**) and its use as
a nucleophile in a substitution reaction. The choice of a solvent is often

| (1) | (2) | (3) |

important. Where these reactions have an ionic character dipolar aprotic solvents, such as dimethylformamide, can be useful. The cleavage (4–5) of simple ethers also requires vigorous conditions (HBr; HI; BBr_3) and involves protonation or coordination to the oxygen atom to facilitate the dealkylation by the nucleophile, which attacks the carbon atom of the ether.[14]

(4)　　　　　　　　　　　　　　　　(5)

Modifications to the reagents involving a combination of a hard acid and a soft base such as aluminium trichloride:sodium iodide in a solvent such as acetonitrile have been used for the cleavage of methyl ethers.[15] The hard Lewis acid ($AlCl_3$) coordinates with the ether oxygen (a hard base) facilitating the attack of the soft iodide nucleophile on the carbon of the methyl ether (6–7).

(6)　　　　　　　　　　Mel　　　　　　(7)

The value of methyl ethers as protecting groups for aliphatic alcohols is limited. This resistance to cleavage provided a means of identifying the free hydroxyl groups in polysaccharides and glycosides in structural studies of these natural products.[16] The position of the methoxyl group in the monomeric sugars obtained after acid-catalysed degradation of the permethylated polysaccharide provided information on the structure of the oligomer. One exception in which methyl ethers have found widespread use is in protecting the hydroxyl groups of acetals in sugars (8–9). The methyl glycosides are formed under mild acidic conditions.

(8)　　　　　H⁺　　　　　　　　　　　CH₃OH　　(9)

The formation of the methyl ether of an alcohol involves a reduction in the extent of hydrogen bonding and hence it can lead to an increase in the volatility of polyhydroxylated compounds facilitating their study by gas chromatography–mass spectrometry (GCMS).

Some reactive alcohols will form an ether via the formation of a carbocation. Montmorillonite KSF clay will catalyse the selective O-alkylation of primary allylic and benzylic alcohols with orthoesters acting as the donor of the alkyl group.[17]

2.2.2 Allyl and benzyl ethers

Changes have been made in the structure of ethers to enhance their ease of formation and cleavage. Allyl and benzyl halides are more reactive than their saturated analogues. Consequently, allyl and benzyl ethers are more readily formed (10–11) and cleaved (11–12) under mildly acidic conditions. Stabilisation of the incipient carbocation by the adjacent

unsaturation facilitates these reactions. Organometallic conditions or, in the case of benzyl ethers, hydrogenolysis, may also be used to cleave these ethers. The unsaturation defines which of the oxygen–carbon bonds is broken thus leaving the carbon–oxygen bond of the protected substrate in place. However, there are sometimes problems in using catalytic hydrogenolysis to cleave benzyl ethers because of the presence of other sensitive functional groups in the molecule. Selectivity can sometimes be achieved by using partially poisoned catalysts, such as palladium on charcoal poisoned with pyridine. The reductive cleavage of benzyl ethers may also be achieved with alkali metals in liquid ammonia or with vigorous reagents such as lithium naphthalenide.[18]

Benzylation can be carried out by using benzyl chloride or bromide and a base such as sodium hydride or sodium hydroxide in a solvent such as dimethylsulfoxide or dimethylformamide.[19] Benzyl ethers have been widely used in sugar chemistry. They are stable to base, to a number of oxidants such as sodium periodate and to reductants such as lithium aluminium hydride. Although they are cleaved under vigorous acidic conditions, they are sufficiently stable to acidic hydrolysis to withstand the conditions that are used to remove isopropylidene and benzylidene groups—a reflection of the ability of the lone pair on an oxygen atom to stabilise an adjacent carbocation in the decomposition of the acetal compared with the effect of an aromatic ring on the benzylic position.

An application of this is found in the preparation of 3-O-benzylglucose (15) by benzylation of 1,2:5,6-di-O-isopropylidene-α-D-glucofuranose (13) to form (14). This was followed by acidic hydrolysis. The 3-benzylglucose (15) was subsequently methylated with methyl iodide and silver oxide in dimethylformamide to give 2,4,6-trimethyl-3-benzyl-α- and β-methyl-D-glucosides (16). Hydrogenolysis of the benzyl protecting

(13)　　　　　　　　　　(14)

(15)　　　　　　　　　　(16)

(17)

group and acid-catalysed hydrolysis of the C-1 acetal methoxyl groups afforded 2,4,6-trimethyl-D-glucose (17).[20] This example reveals not only the differing reactivity of methyl and benzyl ether protecting groups in sugar chemistry but also the use of the isopropylidene group to protect vicinal diols.

2.2.3 Triphenylmethyl ethers

Increasing the number of aromatic groups on the ether carbon of the protecting group increases the reactivity and selectivity of the protecting group by utilising electronic and steric factors. The aromatic rings stabilise adjacent carbocations by delocalisation and provide steric bulk. The sterically demanding triphenylmethyl (trityl) group can be used for the selective protection of primary alcohols. These ethers are prepared from triphenylmethyl chloride and pyridine (18–19) and are cleaved under acidic conditions such as HBr/CH_3CO_2H, $HCl/CHCl_3$ or by

hydrogenolysis. For example it was possible to protect the primary alcohol in (20) whilst leaving the secondary hydroxyl group free for further reactions.

(20)

An application of the differential reactivity of these protecting groups may be seen in the preparation of ketoses from aldoses as exemplified by the conversion of glucose (8) to L-sorbose (26).[21] Treatment of glucose with methanolic hydrochloric acid gave the methyl α-D-glucopyrano-side (21). This was benzylated with benzyl chloride in the presence of potassium hydroxide in dioxan and then the methoxy-acetal was hydrolysed by a mixture of sulfuric acid and acetic acid to give 2,3,4,6-tetra-O-benzyl-D-glucopyranose (22). Reduction of this with lithium aluminium hydride gave the tetrabenzyl-D-glucitol (23) in which the primary alcohol was then selectively protected as the trityl ether (24). Oxidation of the exposed secondary hydroxyl group with dimethyl sulfoxide (DMSO)-acetic anhydride afforded the ketose (25) from which the protecting groups were removed to give L-sorbose (26).

2.2.4   Substituted benzyl ether protecting groups

The ease with which benzyl or trityl ethers can be cleaved may be varied by changing the substituents on the aromatic ring. In the case of

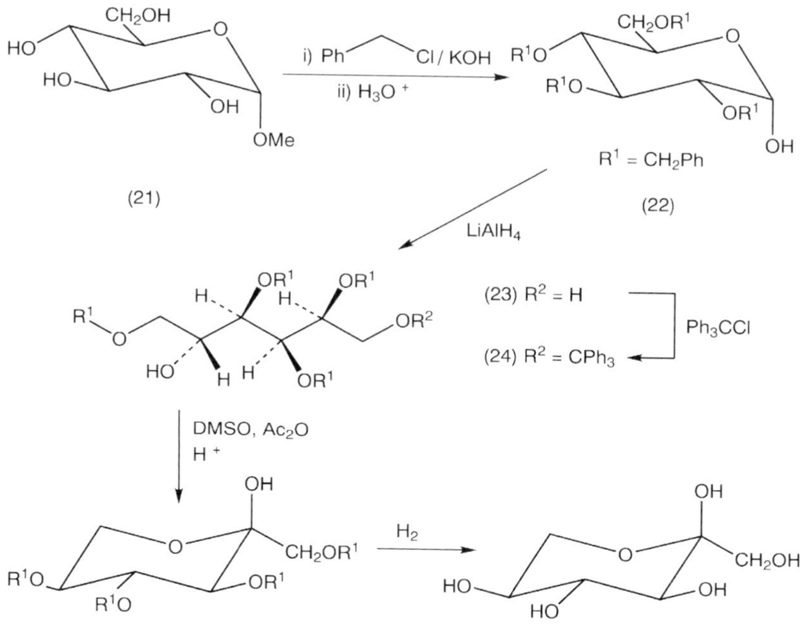

(21)   i) Ph⌒Cl / KOH   ii) H₃O⁺   (22)   R¹ = CH₂Ph

(23) R² = H

(24) R² = CPh₃   Ph₃CCl

LiAlH₄

DMSO, Ac₂O H⁺

(25)   H₂   (26)

oligonucleotides the acidic conditions required for hydrolysis of the trityl group were too drastic and led to decomposition. However, an electron-releasing *para* methoxyl group increased the acid sensitivity of the trityl ether (27) by stabilising the carbocation (28) that is formed during the deprotection. The mono-, di- and tri-methoxy derivatives have a progressively greater ease of hydrolysis.[22]

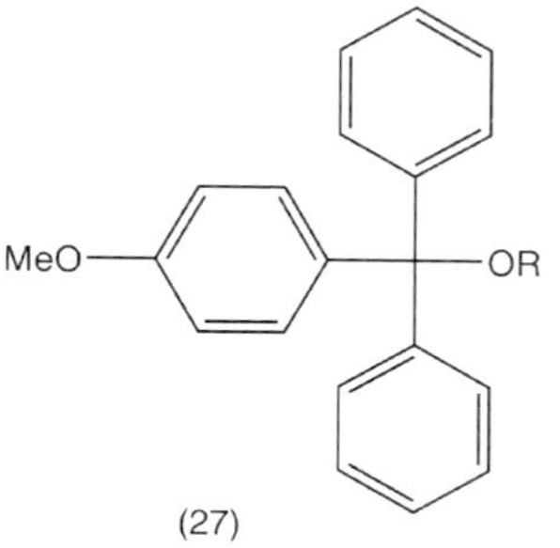

(27)

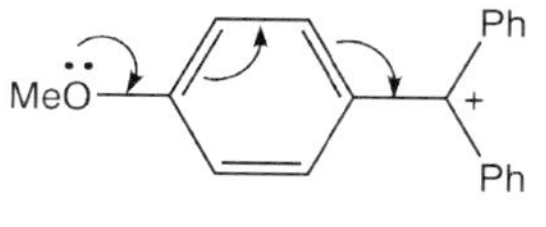

(28)

Several methods for the cleavage of *p*-methoxybenzyl ethers in the presence of unsubstituted benzyl ethers have been developed.

These include oxidation with ceric ammonium nitrate or dichlorodi-cyanobenzoquinone (DDQ) or differential acid-catalysed cleavage with trimethylsilyl chloride and tin(II) chloride, trifluoroacetic acid or magnesium bromide diethyletherate:dimethylsulfide.[23] Iodine in metha-nol has also proved to be an effective reagent for this purpose.[24] Sodium cyanoborohydride and boron trifluoride etherate reductively cleave some p-methoxybenzyl ethers to an easily separable mixture of the correspond-ing alcohol and 4-methylanisole.[25]

It is possible to alter the reactivity of the substituted benzyl ether during the protection–deprotection sequence. Thus the dinitrodiphenyl-methyl ether (29) is relatively stable to acidic conditions.[26] However, once the nitro groups have been reduced to amino groups, the diaminodi-phenylmethyl group (30) may be removed under mildly acidic condi-tions in this example of relay deprotection.

(29)   (30)

Removal of an allyl group may be achieved by using the ease with which an allyl group forms a metal complex with, for example, palladium or titanium.[27] Low-valency titanium will reductively remove a group, and an oxidative procedure using palladium chloride:copper chloride and oxygen has been developed. There are instances in which DDQ will oxidatively remove the allyl ethers of primary alcohols in the presence of secondary allyl ethers.[28]

1,1,1,3,3,3-Hexafluoro-2-phenylisopropyl alcohol (31) reacts selectively

(31)

with primary alcohols under Mitsunobu conditions (diethylazodicarbox-ylate and triphenylphosphine) to give the hexafluoro-2-phenylisopropyl ethers. In this protecting group the electron-withdrawing trifluoromethyl groups oppose the effect of the phenyl group in stabilising an adjacent carbocation. The introduction of the protecting group involves the reagent acting as a nucleophile to displace the phosphorus-activated Mitsunobu intermediate (32) to form the ether (33). These ethers are

(32)   (33)

stable to vigorous reaction conditions but are susceptible to reductive cleavage with lithium naphthalenide.[29] This protecting group may be removed in the presence of trityl, benzyl, tetrahydropyranyl or t-butyl-diphenylsilyl ethers.

2.2.5   Oxygen-substituted ethers

The reactivity of a methyl ether to acid may be enhanced by the addition of a second oxygen atom as in a methoxymethyl ether. The structure of the protecting group is modified to an acetal in which the lone pairs on the oxygen atoms play an important role both in directing the cleavage of a C—O bond and in stabilising a resultant carbocation. Tetrahydro-pyranyl ethers (35) are prepared from the alcohol and dihydropyran (34)[30] in the presence of an acid catalyst, for example toluene-p-sulfonic

(34)   (35)

acid or Amberlyst H-15. The reactivity of the enol ether of dihydropyran is used in the preparation of the acetal, and the additional oxygen atom facilitates the hydrolysis by stabilising a carbocation. Tetrahydropyranyl ethers are stable to strong bases and to most organometallic reagents such as Grignard reagents and lithium alkyls. The free hydroxyl group would react with these reagents. These ethers are also stable to reagents such as lithium aluminium hydride and to acylating and alkylating reagents.

A classic example[31] of the use of the tetrahydropyranyl (THP) protecting group is in the conversion of dehydroisoandrosterone (36) to testosterone (**41**). The 3β-hydroxyl group of dehydroisoandrosterone (**36**) was converted to its THP ether (**37**) and then the C-17 carbonyl group was reduced and the resultant alcohol was protected as its benzoyl ester (**38**). Acid hydrolysis then exposed the 3β-hydroxyl group to give

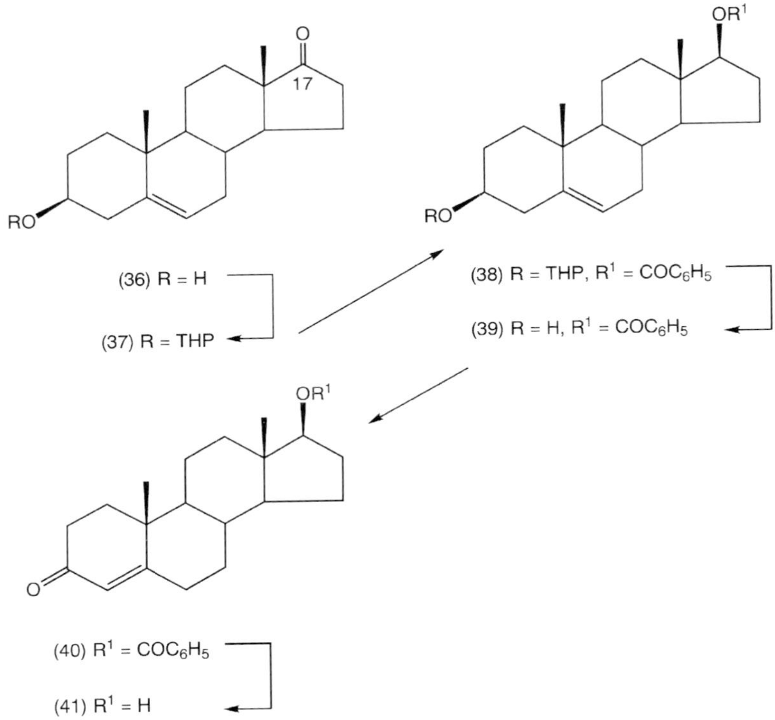
O
17
RO
OR1
RO
(36) R = H
(37) R = THP
(38) R = THP, R1 = COC6H5
(39) R = H, R1 = COC6H5
OR1
O
(40) R1 = COC6H5
(41) R1 = H

(**39**), which was oxidised and the double bond rearranged to afford testosterone benzoate (**40**). Testosterone (**41**) was then obtained by alkaline hydrolysis. This example clearly illustrates the differing reactivity of ether and ester protecting groups. Another simple example[32] involved the protection of the hydroxyl group in propargyl alcohol (**42**). The Grignard reagent (**43**) could then be prepared and carboxylated to afford 4-hydroxybutynoic acid (**44**).

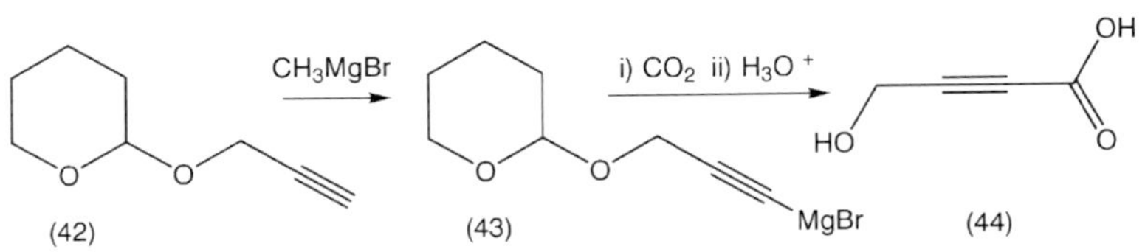
CH3MgBr
i) CO2 ii) H3O +
OH
O
O
O
O
HO
O
MgBr
(42)
(43)
(44)

A wide range of acid catalysts have been used for deprotection to regenerate the parent alcohol. These include protic acids such as hydrochloric acid, acetic acid, toluene-*p*-sulfonic acid and pyridinium toluene-*p*-sulfonate as well as Lewis acids such as magnesium bromide in ether. Molecular sieves and specially prepared forms of graphite are recent additions to this armoury of reagents (e.g. **45 – 46**).[33, 34] The oxidative deprotection of tetrahydropyranyl ethers to form aldehydes and ketones can be carried out with reagents such as dimethylamino-pyridinium (DMAP) chlorochromate in nonaqueous solvents, including acetonitrile(e.g. **45 – 47**).[35]

Reaction of tetrahydropyranylated alcohols with N,N-dimethylphos-gen-iminium chloride (Viehe's salt) yielded[36] the corresponding alkyl chlorides in good yield. Transition metal catalysis using ruthenium(II) or rhodium(III) catalysis with complexes such as [Ru(CH$_3$CN)$_3$ (triphos)] (OTf)$_2$ (**48**) can be used both for introducing and for removing the THP protecting group in acid-sensitive molecules.[37]

One disadvantage of the tetrahydropyranyl ether is that a new chiral centre is generated, with the possibility that mixtures will be formed. This affects both the crystallinity of the derivative and the interpretation of the spectra of the product.

The methoxymethyl protecting group, although not possessing this disadvantage, originally required the use of the toxic chloromethylmethyl ether (49–50) for its introduction. Less hazardous methods using dimethoxymethane (51) and phosphorus pentoxide[38] or trimethylsilyl

(49)   (50)

(51)

iodide[39] catalysis have been introduced. The heterogeneous acid catalyst, Envirocat EP2G® can also be used to catalyse the methoxymethylation of alcohols using dimethoxymethane as the reagent. The methoxymethyl group suffers from the disadvantage that it is rather difficult to cleave, and a number of variants such as the methoxythiomethylgroup[40] have been used which are easier to hydrolyse.

One of the best known of these modified groups is the β-methoxy-ethoxymethyl (MEM) protecting group.[41] This protecting group is introduced by reacting the alcohol with methoxyethoxymethyl chloride (52–53) in the presence of a base such as diisopropylethylamine—a base that is sufficiently hindered to pick up a proton but not to act as a

(52)   (53)

nucleophile. The rationale behind the introduction of this group was the opportunity presented by the oxygen atoms of the ether to participate in bidentate coordination to the metal of a Lewis acid (for example ZnBr$_2$) and thus facilitate the cleavage (54–55). MEM ethers are stable under

(54)

(55)

conditions involving strong bases, hydride reducing agents, organo-metallic reagents and many oxidising agents. Since this grouping is peculiarly sensitive to cleavage by anhydrous zinc bromide or titanium tetrachloride, selective cleavage of the MEM protecting group may be carried out in the presence of acetates, benzoates, benzyl and even tetrahydropyranyl ethers. The combination of the hard acid:soft base pair, trimethylsilyl chloride:sodium iodide in acetonitrile, has provided[42] a useful combination for the cleavage of MEM ethers. Other selective methods of cleavage include modified Lewis acids such as 2-chloro-1,3,2-dithioborolan (56), which have allowed differentiation between silyl and methoxyethoxymethyl groups.

(56)

Neighbouring group participation in the hydrolysis of ether protecting groups has been exploited in several other ways. The guaiacylmethoxy (GUM) protecting group is an example.[43] This protecting group is introduced by using guaiacylmethyl chloride (57–58). Coordination of the

(57)   (58)

guaiacyl moiety to a zinc chloride catalyst (59) plays a role in facilitating

(59)

the cleavage. Another example of assisted hydrolysis is provided by the pentenyloxymethyl (POM) protecting group[44] which is introduced using a chloroether (60–61). Here cleavage of the protecting group is facili-tated by intramolecular cyclisation to a tetrahydrofuran (62–63). This reaction is initiated by the addition of bromonium ion derived from

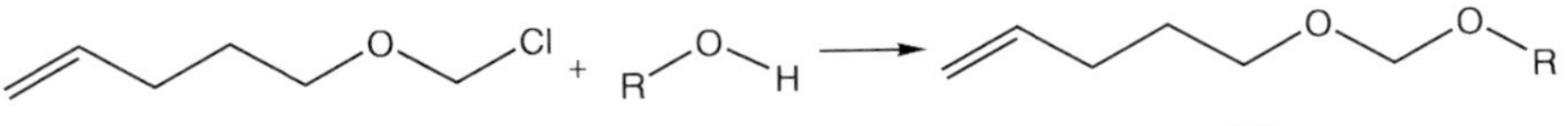

(60)   (61)

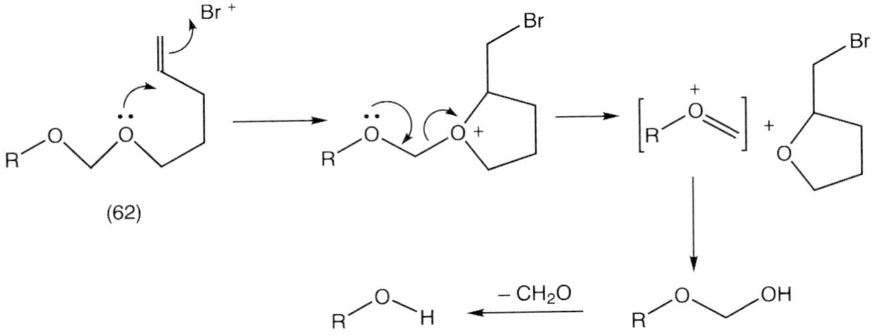

(62)   (63)

N-bromosuccinimide. Another variant on this theme is the *p*-anisoyloxy-methyl group (*p*-AOM) (64)[45] which can be removed oxidatively with ceric ammonium nitrate. The protecting reagent, *p*-anisoyloxymethyl chloride was prepared by the free radical chlorination of 1,4-dimethoxy-benzene (65) in a route which paralleled the preparation of guaiacyl-methyl chloride from veratrole (66).

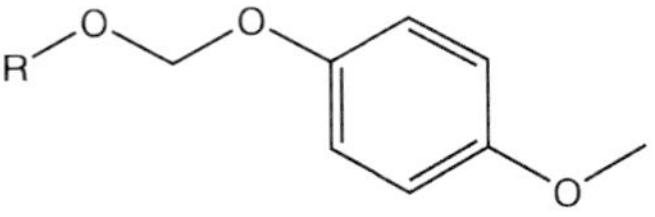

(64)

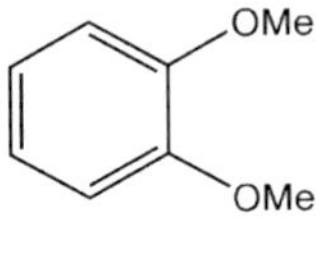

(65)   (66)

Fragmentation reactions also play a useful role in the removal of substituted ether protecting groups. Examples are provided by the reductive cleavage of the trichloroethoxymethyl ether (67) and the fluoride-ion catalysed decomposition of trimethylsilylethoxymethyl ethers (68).[46] The former was used very effectively by Woodward in the

(67)

(68)

synthesis of cephalosporin C.[47] However, these protecting groups have had more use in the protection of carboxylic acids.

2.2.6 Silyl ethers

The discovery of silyl ether protecting groups has revolutionalised the protection of alcohols.[48] The strength of the silicon–oxygen bond provides a driving force for the formation of these ethers. Several methods of cleavage are based on the formation of the particularly strong silicon–fluorine bond. The general method of preparation of silyl ethers (69) involves treatment of the alcohol with a silyl chloride such as trimethylsilyl chloride in the presence of a base. Triethylamine,

(69)

dimethylaminopyridine, DBU (1,8-diazabicyclo[5,4,0]undec-7-ene) and imidazole are commonly employed as bases, and the reactions are carried out in solvents such as dichloromethane. Cleavage is effected by acids such as hydrochloric acid or even acetic acid or by a fluoride ion ($Bu_4N^+F^-$, KF or aqueous HF in CH_3CN). A solution of silicon tetrafluoride in dichloromethane has been used for the cleavage of silyl ethers.[49] The silicon tetrafluoride probably first bonds to the oxygen of the ether and in this way it facilitates the attack of the fluoride on the silicon of the silyl ether. The efficacy of this system is particularly sensitive to steric hindrance by the environment of the oxygen. Carbon tetrabromide in refluxing methanol has been used as a very mild method for desilylation.[50] The reaction may be based on the formation of a small amount of hydrogen bromide.

The trimethylsilyl (TMS) (**70**) group is quite labile and, although it is used to enhance the volatility of hydroxylic compounds for gas chromatography, it is not routinely stable to preparative column chromatography. However, there are a graded series of related protecting groups such as the triethylsilyl (TES) (**71**),[51] t-butyldimethylsilyl (TBDMS) (**72**)[52] and t-butyldiphenylsilyl (TBDPS) (**73**)[53] groups in which

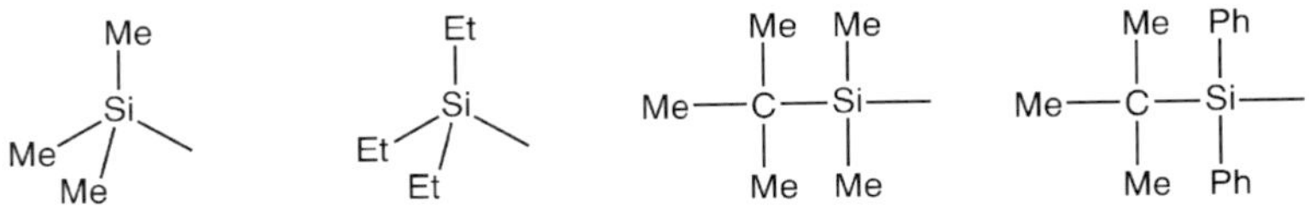

TMS (70)   TES (71)   TBDMS (72)   TBDPS (73)

steric factors increase their stability and diminish their sensitivity to hydrolysis. These steric factors provide the basis for selectivity. The TBDMS and TBDPS groups are particularly sterically demanding. For example, it was possible to protect the less hindered primary hydroxyl group in (**74**) as the TBDMS ether (**75**) and carry out the oxidation of the other primary alcohol in a synthesis of drimenin (**76**). The TES group is

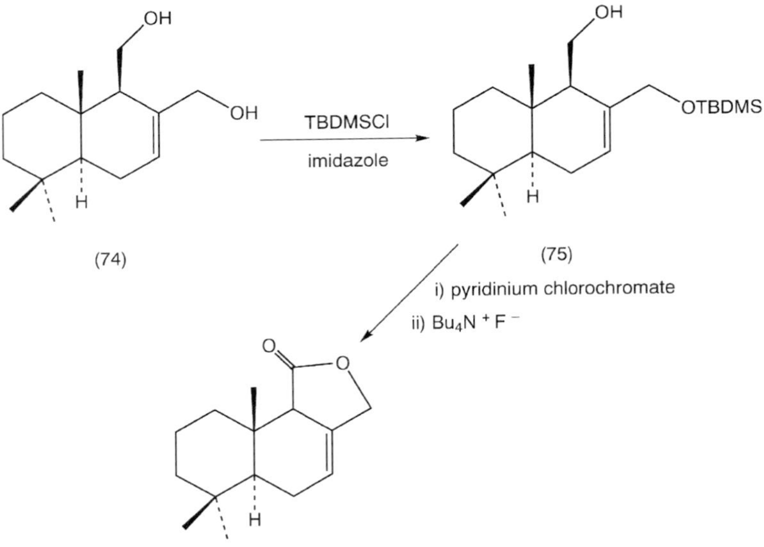

(74)

TBDMSCl
imidazole

(75)

i) pyridinium chlorochromate
ii) Bu$_4$N$^+$F$^-$

(76)

intermediate in stability between the TMS and TBDMS groupings. The triisopropylsilyl (TIPS) ether has a larger steric bulk and greater stability.

The selective hydrolysis of different silyl ethers may be used to distinguish between hydroxyl groups to allow their selective modification. For example in a synthesis of lackcystin analogues, the alcohol (77) was protected with the TBDMS group to afford (78) and then the primary alcohol was exposed in (79) by the selective hydrolysis of the TES group. This example also reveals the use of an oxazoline to protect a vicinal amino alcohol.[54]

Primary allylic and benzylic TBDMS ethers are more sensitive and can be selectively cleaved in hot aqueous dimethylsulfoxide. The t-butoxy-diphenylsilyl protecting group has a comparable stability to the t-butyldiphenylsilyl group. It is, however, more easily prepared from diphenyldichlorosilane.[55] The t-butoxydiphenylsilyl group can be cleaved with sodium sulfide.[56]

The value of the bulky t-butyldimethylsilyl protecting group as a directing group may be illustrated by the following example. Epoxidation of the alkene in (80) by m-chloroperbenzoic acid was directed to the β face by hydrogen bonding from the adjacent 3β-hydroxyl group to give (81). On the other hand, epoxidation of the protected allylic alcohol (82) afforded the α-oriented epoxide (83). The protecting group was then removed with tetrabutylammonium fluoride to give the 3β-hydroxy-4α,5α-epoxide (84) without cleaving the epoxide.

Polysilanes, exemplified by tris(trimethylsilyl)silane, absorb light in the ultraviolet (UV) region and fragment to form silylenes. Tris(trimethylsilyl)ethers (sisyl ethers) are readily prepared (85–86) from primary and secondary alcohols and, because of their bulk, they are relatively stable. They are, however, unstable to mild photolytic conditions, thus representing a novel addition to the range of silicon protecting groups.[57]

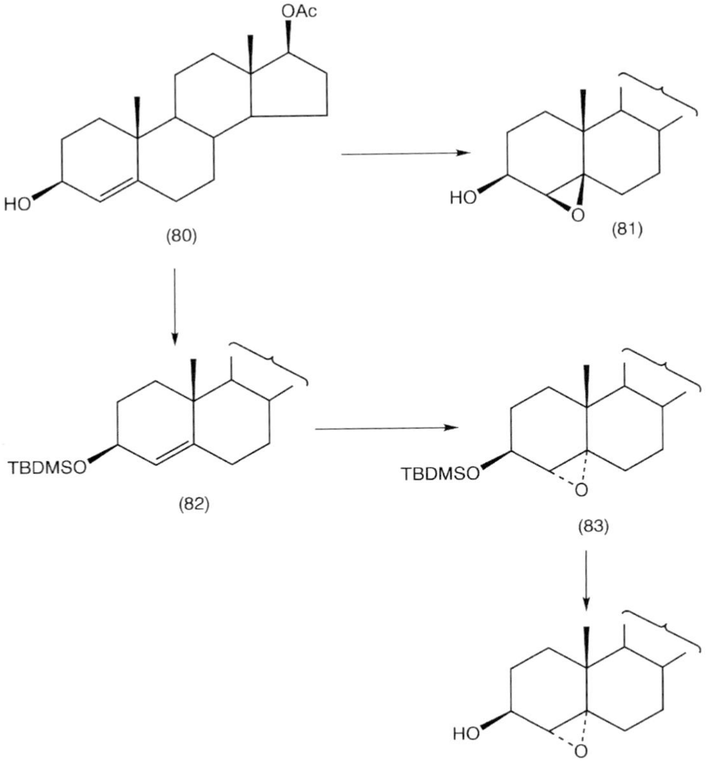

(80)

(81)

(82)

(83)

(84)

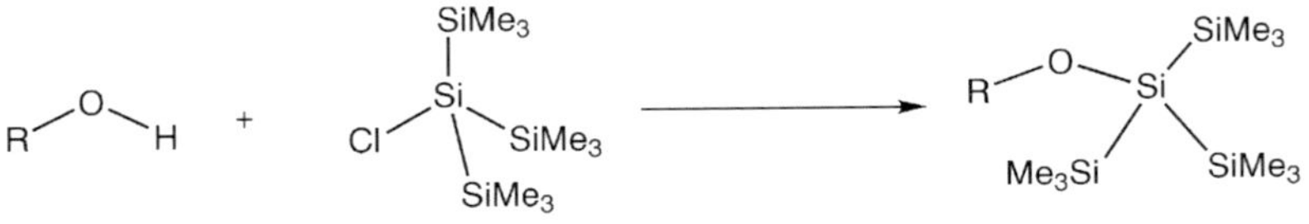

(85)

(86)

2.3   Esters as protecting groups for alcohols

Hydroxyl groups can be protected as ethers for reactions under basic conditions and as esters for reactions under acidic conditions. The interaction of the electron-withdrawing carbonyl group with the lone pairs of the ester oxygen renders this atom less basic and hence less acid sensitive. There are a range of acylating groups which have been used. These show varying selectivities and require different methods for the removal of the protecting group.

2.3.1 Acetate esters

Acetate esters are the most frequently used ester protecting groups. There are a range of conditions for introducing the acetate group. These include, in order of increasing vigour: acetic anhydride:pyridine; acetic anhydride:pyridine:dimethylaminopyridine; acetic anhydride:sodium acetate; acetic anhydride:toluene-*p*-sulfonic acid; acetyl chloride:pyridine; and ketene. Dimethylaminopyridine (**87**) is a useful acylation catalyst because of the formation of highly reactive N-acylpyridinium salts (**88**)

(87)

(88)

from the anhydride.[58] The use of a sterically hindered base such as N,N-diisopropylethylamine with acetyl chloride permits the selective acetylation of a primary alcohol in the presence of a secondary alcohol.[59] Although the acylation of alcohols by acid anhydrides or acyl chlorides is often carried out in the presence of a base, such as a tertiary amine, acid catalysts may be used. Scandium trifluoromethanesulfonate[60] is a very efficient Lewis acid catalyst for the acylation of alcohols with acid anhydrides. The Montmorillonite clays K-10 and KSF have also been found[61] to be useful solid acid catalysts for the acetylation of alcohols with acetic anhydride in dichloromethane.

Even though many hydroxyl groups are readily acetylated it is possible to devise conditions which allow selectivity. Thus steric factors make the rates of acetylation of steroidal 3β-, 6β- and 11β-hydroxyl groups (**89**)

(89)

sufficiently different to allow a distinction to be made between them.[62]

Similarly, there is sufficient difference in the rate of hydrolysis of the
3β- and 17β-acetates for it to be possible to hydrolyse selectively a 3β,
17β-diacetate (90) to obtain the 17β-acetoxy-3β-alcohol (91). The use of a
bulky π acid catalyst such as tetracyanoethylene in methanol can enhance
the effect of steric factors.[63]

aq. K$_2$CO$_3$

(90)   (91)

2.3.2 Modified esters

It is possible to introduce greater selectivity into ester formation and
hydrolysis by varying the structure of the esterifying acid. Formate esters,
which can be introduced by means of a formic acid:acetic anhydride
mixture,[64] can be hydrolysed by using mildly basic reagents such as
potassium bicarbonate in aqueous methanol. Trifluoroacetyl esters (92)
are also formed and hydrolysed easily whereas chloroacetate, methoxy-
acetate and phenoxyacetate esters (93–95, respectively) have also been
used in situations where a mild hydrolysis is required.[65] The benzoyl (96)
grouping, however, possesses a greater selectivity both in its formation
and in its removal. The same applies to the bulky trimethylacetate
(pivaloate) esters (97). For example whereas a steroidal 17β-acetate is
often partially hydrolysed during the alkaline conditions necessary for the
oxidation of a borane in hydroboration, the 17β-pivaloate is stable.

(92)   (93)   (94)

(95)   (96)   (97)

One problem with the acetate protecting group is that of acyl migration and acetoxylinium ion participation in reactions. For example treatment of the acetate (98) with mild base leads via (99) to a mixture of (98) and (100). In the allylic substitution of the 3β-acetate (101) at C-4, by using bromine and silver acetate, the product is (102), which probably arises via

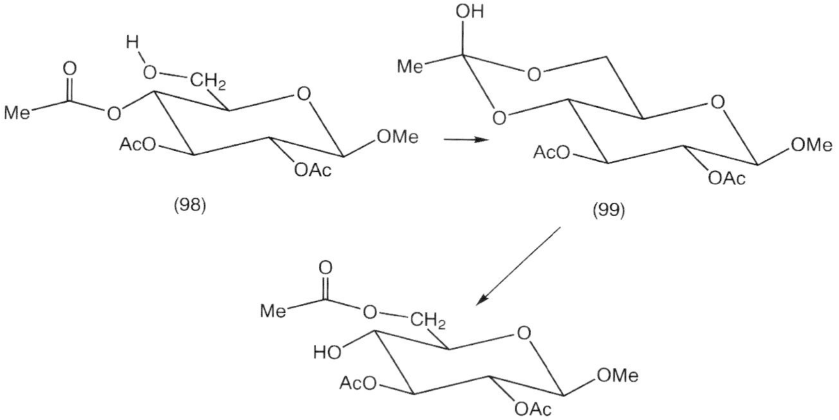

(98)

(99)

(100)

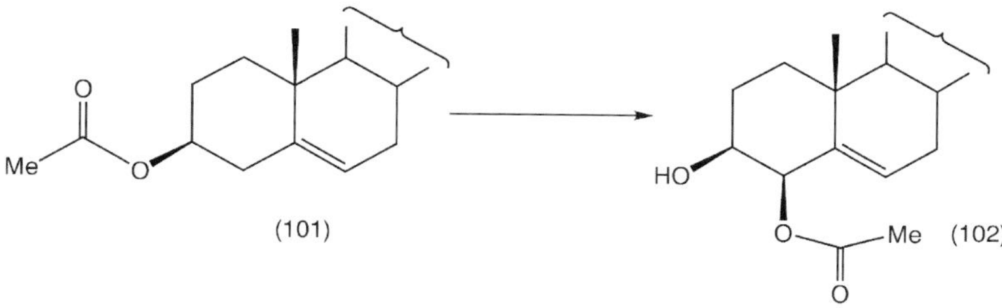

(101)

(102)

(103). In another example epoxidation of the alkenes (104) and (105) affords a similar mixture of (106) and (107).[66] Benzoate groups show less of a tendency to migrate.

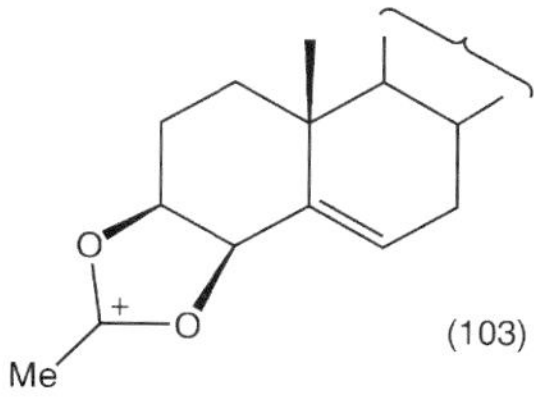

(103)

Neighbouring group participation can play a useful role in the hydrolysis of esters. As an example, diphenylacetyl chloride (108) is sufficiently bulky to protect selectively the primary alcohol of sugars.[67]

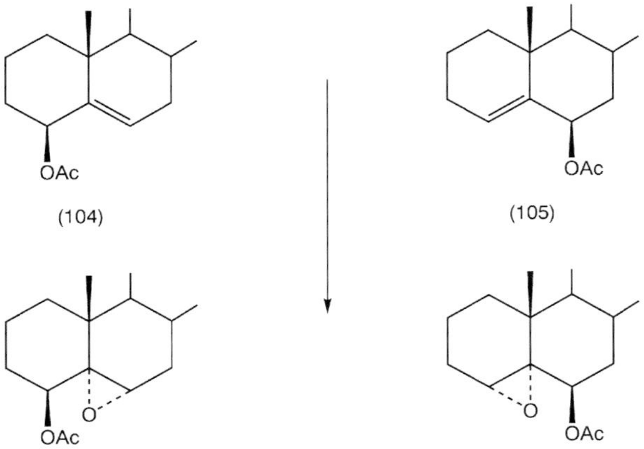

(104)

(105)

(106)

(107)

A mild method has been developed for removing this protecting group to avoid disrupting the glycosyl bond of nucleosides. The benzylic C—H bond of (109) is readily brominated by N-bromosuccinimide (NBS) to give (110). The protecting group is then removed by using thiourea, as in (111). The nucleophilic sulfur of the thiourea displaces the benzylic

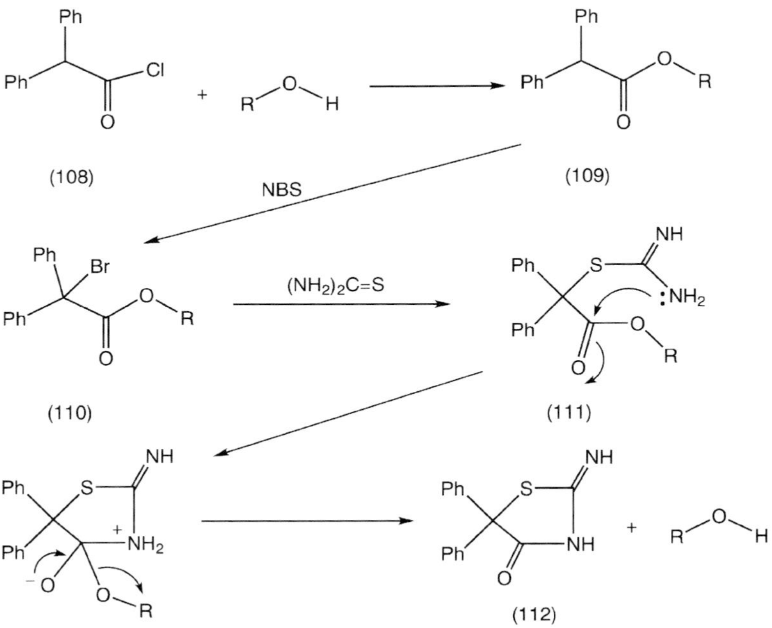

(108)

(109)

NBS

$(NH_2)_2C=S$

(110)

(111)

(112)

halogen and allows the —NH$_2$ group to come into close contact with the carbonyl group of the ester, facilitating the hydrolysis (111–112).

The additional oxygen atom in carbonate esters, such as the ethoxy-carbonyl (cathylate) group (113), makes them more stable to base than the corresponding acetates.[68] These esters are prepared by reaction of the alcohol with ethyl chloroformate and pyridine. Selectivity in their formation based on stereochemical differences has been observed. For example the cholestane-3β, 5α, 6α-triol (114), possessing an equatorial 6-hydroxyl group, gave a 3, 6-diester (115). However, the corresponding axial alcohol, cholestane-3β, 5α, 6β-triol (116), gave only a 3β-derivative (117). In neither case was the tertiary 5α-alcohol esterified.

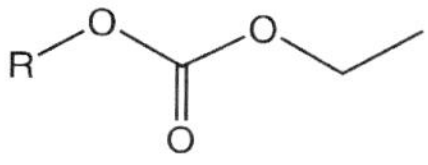

(113)

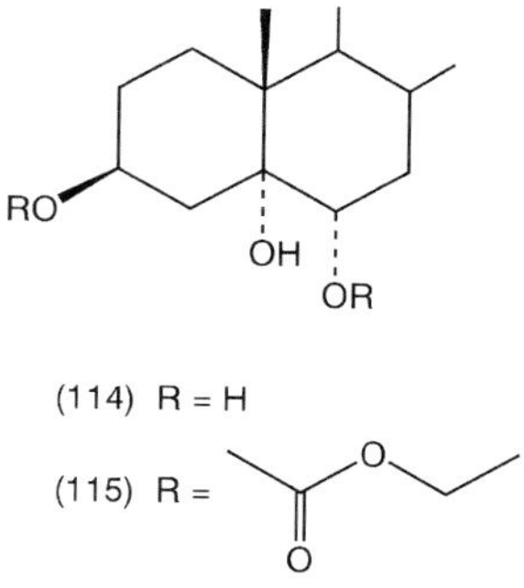

(114) R = H

(115) R =

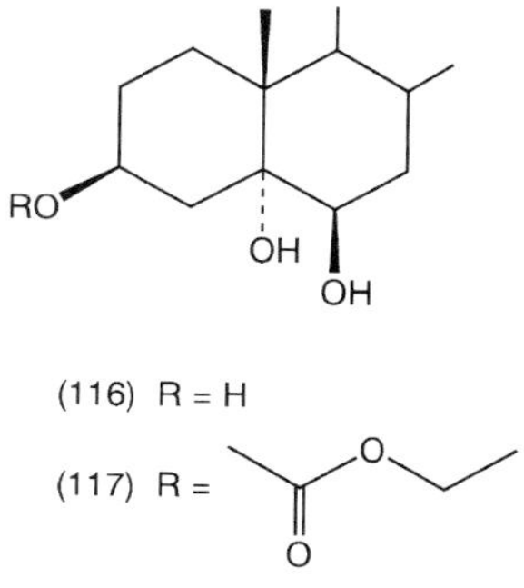

(116) R = H

(117) R =

Modification of a carbonate ester to give, for example, the β-trichloroethylcarbonate (118) permits alternative reductive methods to be used for the removal of the protecting group.[69] A fragmentation

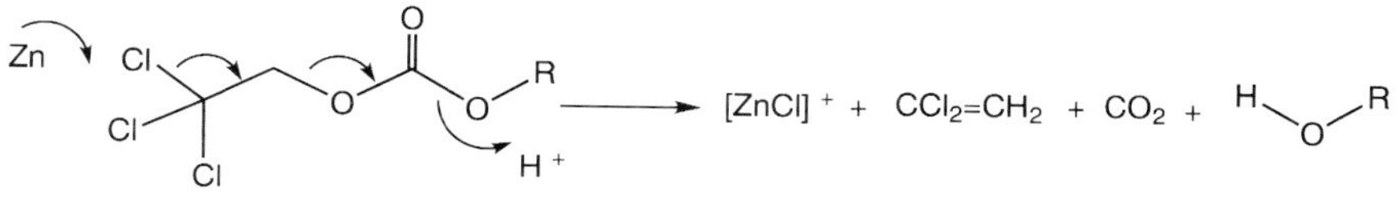

[ZnCl]$^+$ + CCl$_2$=CH$_2$ + CO$_2$ +

(118)

reaction based on a β-trimethylsilyl group (119) has been used in a similar fashion.

(119)

2.4   The protection of 1,2-diols and 1,3-diols

Simple treatment of a symmetrical diol with a stoichiometric amount of the protecting reagent can lead to a mixture of unreacted diol, mono-protected and bis-protected products. However, conditions have been developed for the monobenzylation of symmetrical diols.[70] Although it is possible to protect the hydroxyl groups of unsymmetrical vicinal diols separately, there are advantages in using a cyclic system involving both hydroxyl groups. Vicinal axial:equatorial and diequatorial diols and some 1:3-diols can be protected as cyclic acetals by condensation with formaldehyde, acetaldehyde, benzaldehyde or acetone in the presence of a catalyst such as toluene-p-sulfonic acid or anhydrous copper sulfate. However, rigid *trans* diaxial 1,2-diols have to be protected separately.

There is an interesting dichotomy in carbohydrate chemistry. The condensation of glucose (**120**) with benzaldehyde leads to the protection of a 1:3-glycol and the stabilisation of the pyranose form (**121**) whereas acetone forms the bisacetonide (**122**) of the furanose form. These

derivatives are both stable under basic conditions. A typical example of their use involved the 4,6-ethylidene derivative of glucose (**123**), which was prepared from glucose by reaction with the acetaldehyde trimer, paraldehyde and sulfuric acid. It was oxidised with sodium periodate to the 2,4-ethylidene derivative of erythrose (124). Reduction with sodium borohydride and acidic hydrolysis gave erythritol (125).[71]

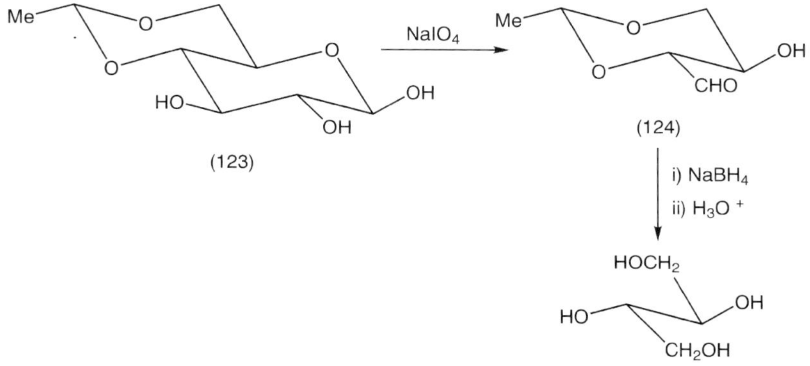

(123) (124) (125)

The selective protection of the 1,3-hydroxyl groups of glycerol (126) as a benzylidene derivative (127) allowed the esterification of the secondary alcohol.[72]

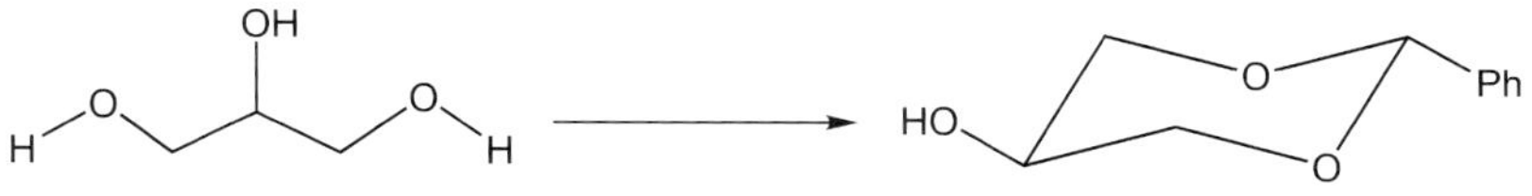

(126) (127)

Acetonides (129) form useful protecting groups since, unlike the ethylidene and benzylidene groups, the protecting group does not introduce a new chiral centre. They are prepared from the glycol (128) and acetone in the presence of an acid catalyst or by an acetal exchange

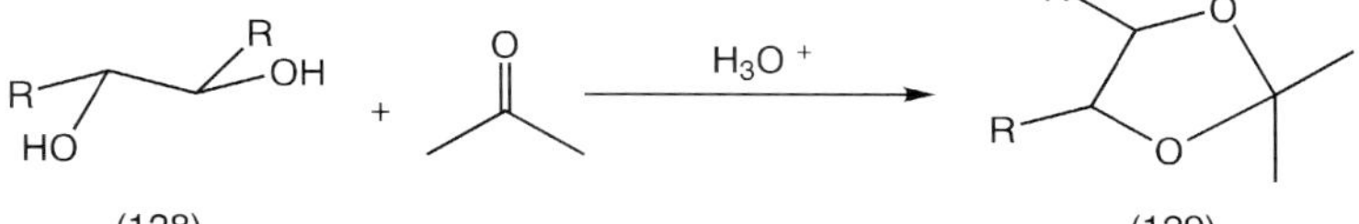

(128) (129)

reaction from 2,2-dimethoxypropane (130). The highly reactive enol ether, 2-methoxypropene can also be used to form acetonides. The lone pair on the oxygen stabilises the reactive intermediates (131). An example

(130)   (131)

of the use of an acetonide is in the preparation of monoglycerides (133) from glycerol (132). The formation of a 1,2-acetonide contrasted to the formation of the 1,3-benzylidene derivative (127).[72]

(132)   (133)

The use of dispiroketals (dispoke derivatives) for protecting 1,2-diols has attracted interest.[73] The spiroketals [for example, 136 from S-(-)-1,2, 4-butanetriol (134)] are introduced by acid-catalysed addition to 6,6'-bisdihydropyran (135). The conformation of these protected units is

(134)   (136)

determined by the anomeric effects of the lone pairs on the oxygen atoms and of the tendency of substituents where possible to take up an equatorial configuration. These protecting groups have a tendency to react preferentially with *trans* diequatorial diols. For example methyl α-D-galactopyranoside (137) gives the 2,3-protected derivative (138), allowing modification of the other hydroxyl groups. Cyclohexane-1,2-diacetals derived from 1,1,2,2-tetramethoxycyclohexane have also been

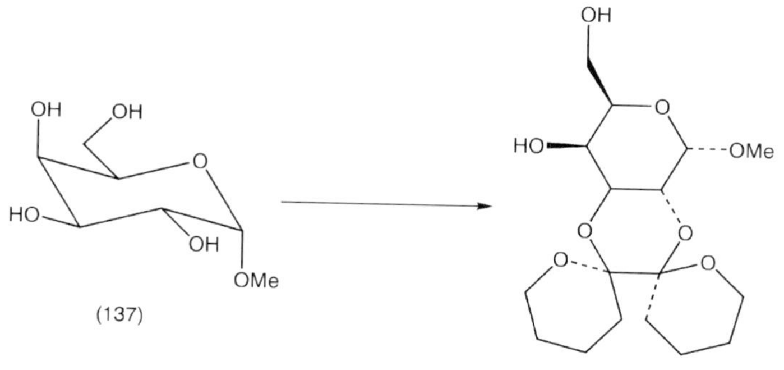

(137)

(138)

investigated[74] as protecting agents for diequatorial vicinal diols. These may have some advantages over the dispoke derivatives.

The silyl ether (139) represents a silicon equivalent of the acetonide grouping. There are also a number of protecting groups involving boron in the ring system.

Whereas the cyclic acetals are stable to base and are labile under acidic conditions, cyclic carbonates as esters are relatively stable to acid. They are prepared by the action of phosgene (140) or the less toxic

(139)

$COCl_2$

(140)

bistrichloromethylcarbonate (triphosgene) (141) in pyridine.[75, 76] Glucose for example affords the 1,2:5,6-dicarbonate (142). An example of the use

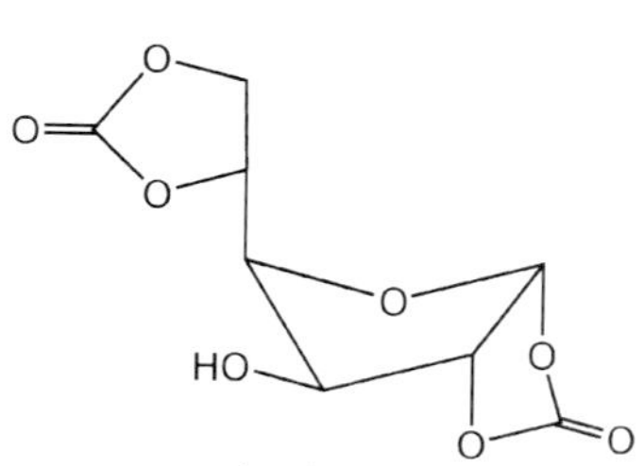

(142)

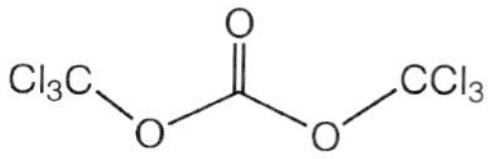

(141)

of a cyclic carbonate in synthesis involves the conversion of methyl β-D-glucofuranoside-5,6-carbonate (143) via (144) to D-erythrose (145).[77]

(143)   (144)   (145)

Adsorption of an unsymmetrical diol on alumina may take place predominantly at the less hindered alcohol. This can allow selective reaction at the more sterically hindered, nonadsorbed, site. Acetylation with acetyl chloride then showed selectivity in forming the more hindered secondary acetate.[78]

The thionocarbonate protecting group can be introduced by reacting a 1,2-diol with 1,1'-thionocarbonyldiimidazole. Thionocarbonates possess similar properties to carbonates, being stable in acid, but are hydrolysed more easily by base. However, the presence of the sulfur confers additional properties. For example the sulfur may be hydrogenolysed with triphenyltin hydride to afford the methyleneacetal or with tributyltin hydride to afford the alkene. In a polyol ring closure on 1,2-diols is preferred over alternative 1,3-diol functionality.[79]

Orthoesters provide the opportunity for the simultaneous protection of three hydroxyl groups.[60-82] This method of protection has been applied in the synthesis of D-myoinositol-1,4,5-triphosphate. Inositol (146) was heated with ethyl-orthoacetate in the presence of toluene-p-sulfonic acid as a catalyst to give (147). A consequence of this particular protection was that the axial and equatorial relationships of the remaining free

(146)   (147)

hydroxyl groups was reversed compared with the free myo-inositol. The protected compound was regioselectively silylated and benzoylated at the equatorial hydroxyl group.

2.5   The protection of phenols

The protection of phenols follows the pattern of the protection of alcohols but with some difference in selectivity arising from the acidity of the phenolic hydroxyl group. It is often important to protect a phenol not only because of the acidity of the phenol but also because of the tendency of phenols to undergo oxidation reactions.

Although methyl ethers are readily prepared with dimethylsulfate or with diazomethane[83] they are often difficult to cleave even when reagents such as boron tribromide are used.[84] A complex between boron tribromide and dimethylsulfide is used, with the dimethylsulfide providing a soft nucleophile to attack the methoxyl carbon.

Phenolic hydroxyl groups which form part of strongly chelated *ortho*-hydroxycarbonyl or perihydroxycarbonyl systems (for example **148**)[85] are

(148)

not easily methylated by diazomethane. Similar differences in reactivity are found in methylation with dimethylsulfate and alkali. For example methylation of (**149**) gave (**150**). The selectivity may be modified by methylation of polyhydroxylic phenols in the presence of borate salts

(149)   (150)

which form complexes with catechols. Thus gallic acid (**151**) may be partially methylated to form (**152**).[86] The methylenation of catechols may be achieved by using dichloromethane and a low concentration of the dianion in a polar aprotic solvent.[87] This can avoid the formation of

(151)   (152)

dimeric species in which the methylene group forms a bridge between two molecules.

Aromatic substituents vary the ease of hydrolysis of phenolic ethers. For example alkaline hydrolysis of 4-nitroveratrole (153) leads to (154), whereas under acidic conditions the product is (155). The former arises by

nucleophilic addition directed by the nitro compound. Phenyl ethers may be cleaved by dry lithium iodide or magnesium iodide. The role of the metal is to coordinate to the oxygen and to facilitate attack by the nucleophile. Other reagents such as boron tribromide have been used but there are occasional reports of substitution of reactive aromatic rings. In this context a useful protecting group for phenols is the benzyl ether since it can be removed by hydrogenolysis.

3 The protection of aldehydes and ketones

3.1 Introduction

The reactivity of a carbonyl group towards nucleophiles, its ability to render an adjacent C—H acidic and its tautomeric relationship with an electron-rich enol are properties which are valuable in synthesis. However, these can lead to complications in the synthesis of polyfunctional molecules. Consequently, there is often a need to mask the reactivity of a carbonyl group during the course of a synthesis in order to manipulate another functional group.

3.2 Acetals as protecting groups[88]

One of the commonest protecting groups for a carbonyl is the dimethoxyacetal (1) which is prepared from an aldehyde or an unhindered ketone by reaction with methanol in the presence of an acid catalyst. The equilibrium involves the generation of water and can be

$$
\underset{}{\text{O}} + \text{MeOH} \;\underset{}{\overset{\text{H}_3\text{O}^+}{\rightleftharpoons}}\; \underset{}{\text{MeO} \quad \text{OH}} + \text{MeOH} \;\underset{}{\overset{\text{H}_3\text{O}^+}{\rightleftharpoons}}\; \underset{}{\text{MeO} \quad \text{OMe}} + \text{H}_2\text{O}
$$

(1)

moved to the right not only by the use of excess alcohol but also by the use of a water scavenger such as trimethyl or triethyl orthoformate or by the azeotropic removal of the water.[89] Acetals are easily hydrolysed with acid via a reactive electrophilic intermediate (2). Reaction of dimethoxymethane with toluene-p-sulfonic acid or with phosphorus oxychloride affords a useful methylenation reagent.

$$
\underset{}{\text{MeO} \quad \text{OMe}} \;\underset{}{\overset{\text{H}_3\text{O}^+}{\rightleftharpoons}}\; \underset{}{\overset{+\,\text{O}\!-\!\text{Me}}{}} + \text{MeOH}
$$

(2)

An example of the selective formation of a dimethoxyacetal is illustrated by a step in the synthesis of a tetrahydroquinazoline analogue of folic acid.[90] Treatment of the Diels–Alder adduct (3) with methanolic ammonium chloride gave the dimethoxyacetal (4) of the aldehyde at the same time as hydrolysing the enol ether to the ketone where further reaction to give (5), and thence (6), was possible. A further example[91] of

this selectivity is shown in the degradation of cholesterol (7) to provide a synthon for rings C and D of vitamin D. Ozonolysis of cholesteryl acetate (8) gave the keto-aldehyde (9). The aldehyde was converted to the dimethoxyacetal (10) with methyl orthoformate in methanol containing toluene-*p*-sulfonic acid as a catalyst while the ketone remained unchanged. The protecting group survived intact throughout the elimination of the acetate with methanolic sodium methoxide to form (11) and the photochemical degradation of ring A to give (12). The protecting group was hydrolysed at the end of the reaction sequence with aqueous hydrochloric acid in tetrahydrofuran. Alumina has been used[92] as a catalyst for the selective acetalisation of aldehydes in the presence of ketones.

An important area where the methoxyl group plays a useful role as a protecting group is in the protection of the cyclic hemiacetals of sugars. These methoxyacetals are formed by treatment with methanolic hydrochloric acid. This protecting group is often introduced as the first step in a sequence involving the manipulation of a series of protecting groups in order eventually to leave exposed a specific hydroxyl group.

3.3   1,3-Dioxolanes and related protecting groups

The 1,3-dioxolanes (13) are widely used as protecting groups for aldehydes and ketones.[93] The formation of a ring system imparts an

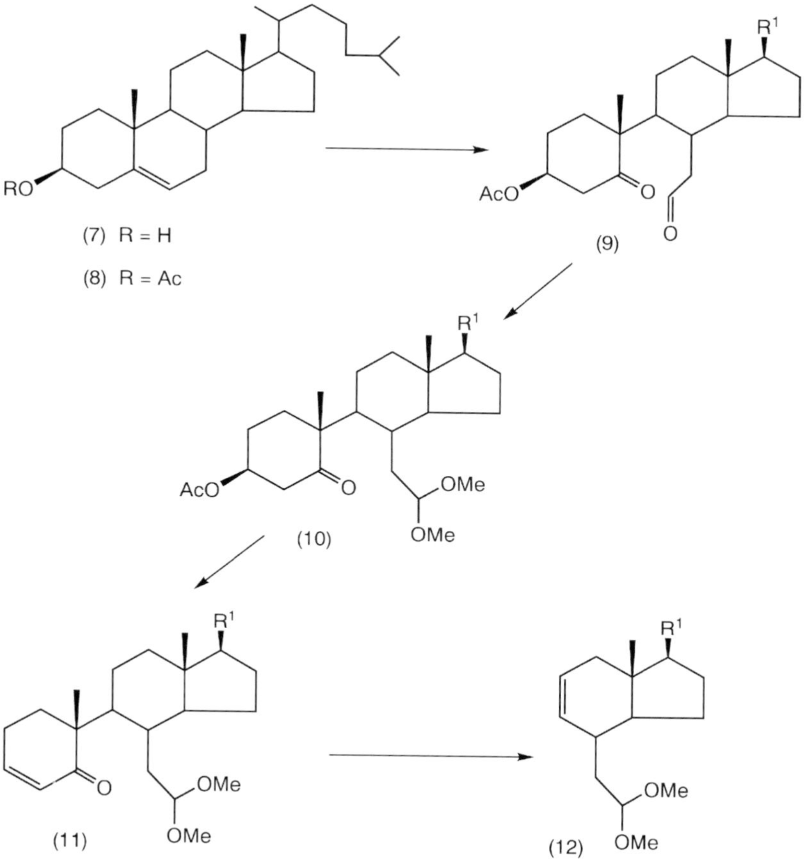

(7) R = H

(8) R = Ac

(9)

(10)

(11)

(12)

extra stability to the group. The 1,3-dioxolanes are formed by the treatment of an aldehyde or an unhindered ketone with ethylene glycol in the presence of an acid catalyst such as toluene-p-sulfonic acid. An alternative method is to use an exchange reaction in which the donor

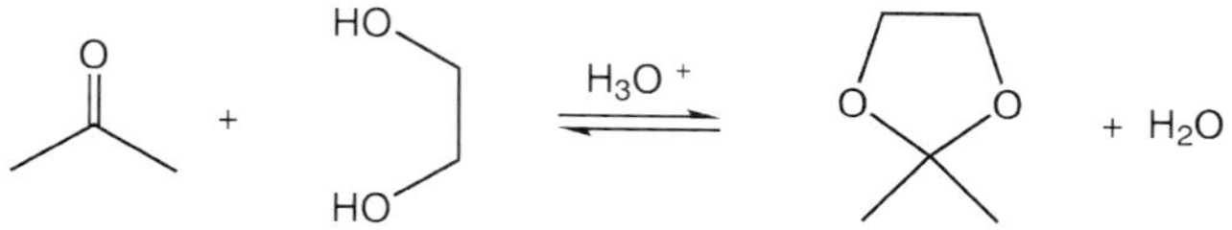

(13)

molecule is the dioxolane of a low boiling ketone such as acetone or butanone.[94] The reaction is carried out under reflux so that the acetone or butanone is removed by distillation as it is formed. An alternative exchange reaction utilises the ethylene acetal of dimethylformamide (14) as the donor molecule.[95] In these exchange reactions the lone pairs on oxygen or nitrogen are important in stabilising a reactive intermediate (15) thereby facilitating the transfer of the diol to afford (13).

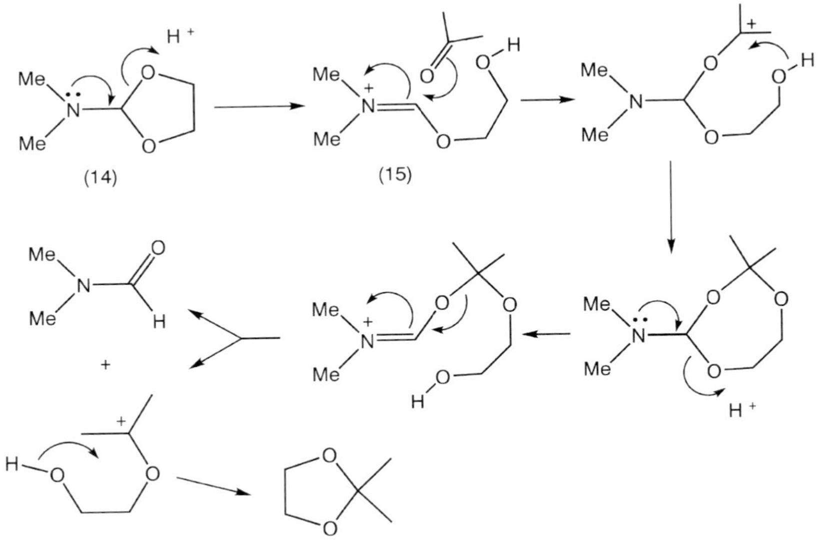

(14)   (15)

(13)

Steric factors confer selectivity on the ease of preparation of 1,3-dioxolanes with the approximate order: aldehydes > acyclic ketones and cyclic hexanones > cyclopentanones > αβ-unsaturated ketones > aryl ketones. However, this order can be modified significantly by structural features in the environment of the carbonyl group. In the steroids (e.g. 16),

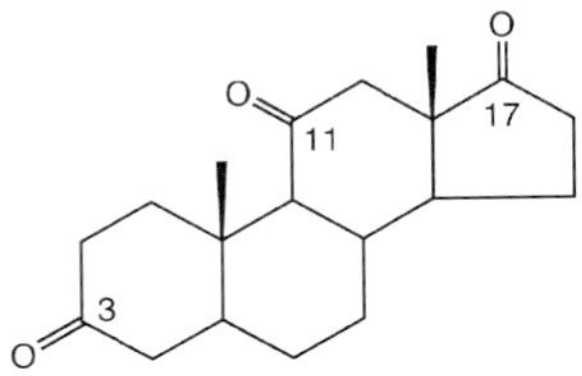

(16)

the relatively unhindered C-3 ketone reacts more easily than the C-17 ketone which in turn reacts more easily than C-12 and C-20. The sterically hindered C-11 ketone is relatively inert. With the unsaturated ketone (17) the double bond moved out of conjugation to the Δ^5-position (18).

This deconjugation was used in the synthesis of the sesquiterpenoid, bulnesene.[62] The selectivity in acetal formation arising from steric factors allowed the preparation of the monoacetal (20) of the diketone (19) and the subsequent use of the protected product in carotenoid synthesis.[96]

Dioxolanes are stable to alkaline, neutral and some organometallic reagents. They are unaffected by many of the acylating reagents (acetic anhydride, acetyl chloride, toluene-p-sulfonyl chloride), oxidising, dehydrating and chlorinating reagents that attack alcohols and by the oxidising reagents that attack alkenes, including osmium tetraoxide and peracids. However, vigorous organometallic reactions which utilise an oxyphilic Lewis acid such as an aluminium, boron or titanium halide can result in their cleavage.

Dioxolanes are cleaved by treatment with acid, iodotrichlorosilane[97] or boron trifluoride etherate and the iodide ion.[98] The hard silicon or boron attacks the oxygen, and the soft iodide ion reacts with the carbon. Oxidative methods of cleavage based on the sensitivity of ethers to peroxidative oxidation have also been described.[99,100]

There are situations in which other acid-sensitive groups are present in the molecule and milder methods of cleavage are required. Various modifications of the dioxolane have been introduced to facilitate this. Protection of a ketone by conversion to the 5,5-dibromo-1,3-dioxolane using 2,2-dibromopropane-1,3-diol (21) in the presence of toluene-p-sulfonic acid provides an interesting method[101] as exemplified by the conversion of (22) to (23). This protecting group does not introduce any new chirality and it protects the carbonyl group against reagents such as sodium borohydride and peracid. It is removed reductively with a zinc:silver couple in acetic acid (24–25). Another protecting group of this kind is the bromomethylethylene acetal (27)[102] formed by reaction with

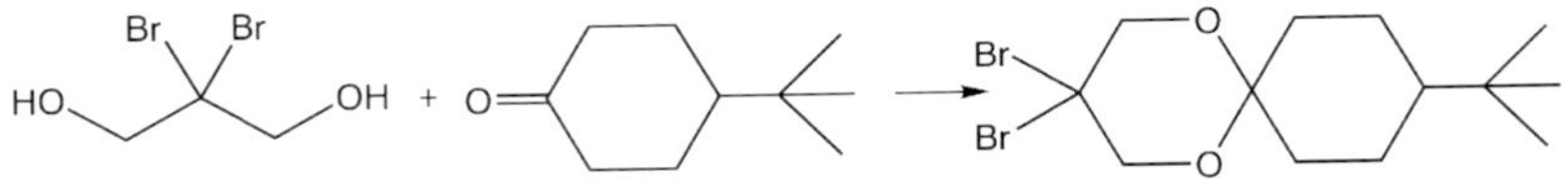

(21)   (22)   (23)

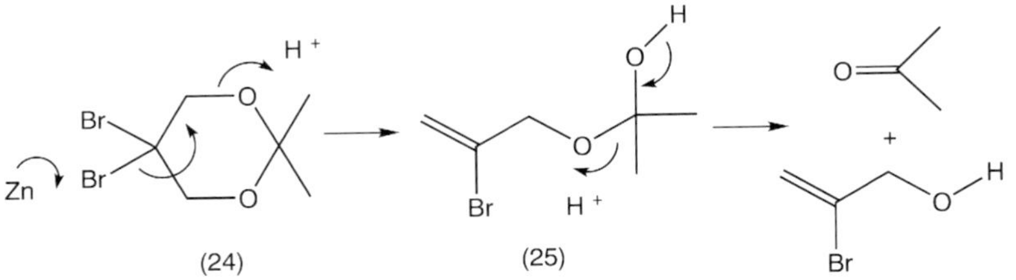

(24)   (25)

the bromoglycol, 1,2-dihydroxy-3-bromopropane (**26**), which is readily available. The regeneration step involves treatment of the acetal with zinc

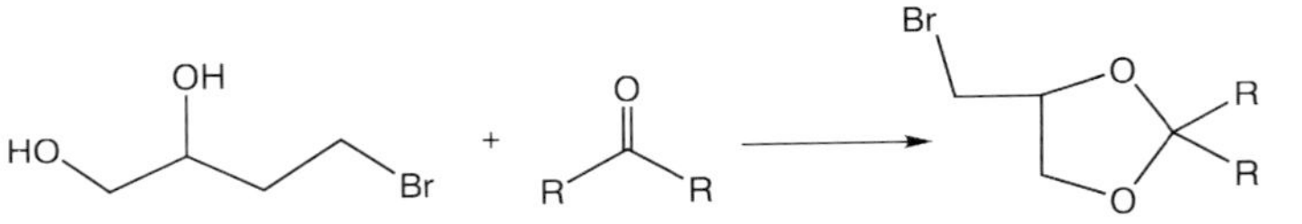

(26)   (27)

to bring about a reductive elimination (**28**). A mechanistically similar fragmentation utilises the trimethylsilylmethyl-1,3-dioxolane protecting

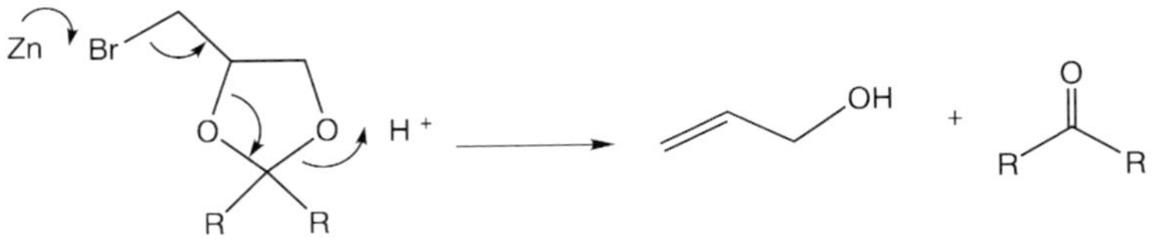

(28)

group (**29**)[103] where the deprotection step is based on the affinity of the fluoride ion for silicon. The paramethoxybenzyl moiety is known to be susceptible to benzylic oxidation and use has been made of this in the

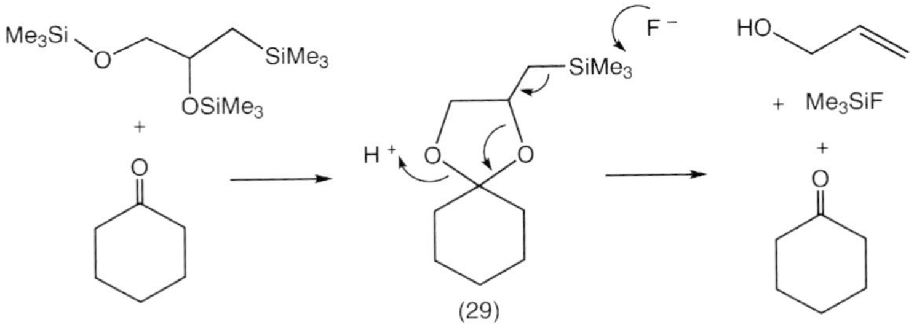

(29)

selective cleavage of dioxolanes, for example in the regeneration of (31) from (30).[104]

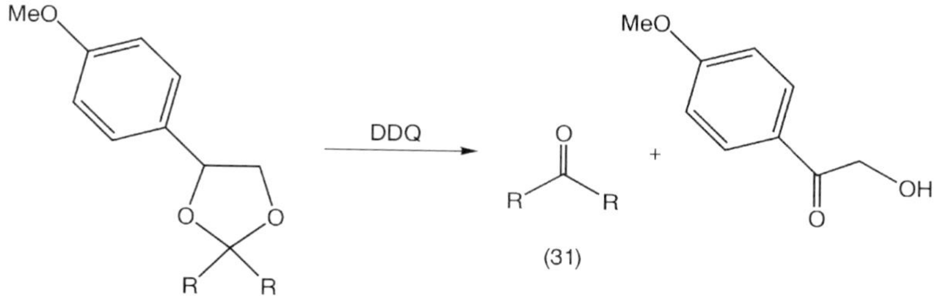

(30) (31)

Aldehydes are normally more reactive than ketones but there are occasions when it is necessary to protect a ketone and leave an aldehyde exposed for reaction. This has been achieved[105] by the selective conversion of the aldehyde to a 1-silyloxysulfonium ion (32) by trimethylsilyl triflate and dimethylsulfide. The dioxonolation of the ketone with 1,2-bis(tri-methylsilyloxy)ethane followed by an aqueous hydrolysis of the sulfonium salt regenerates the aldehyde (33).

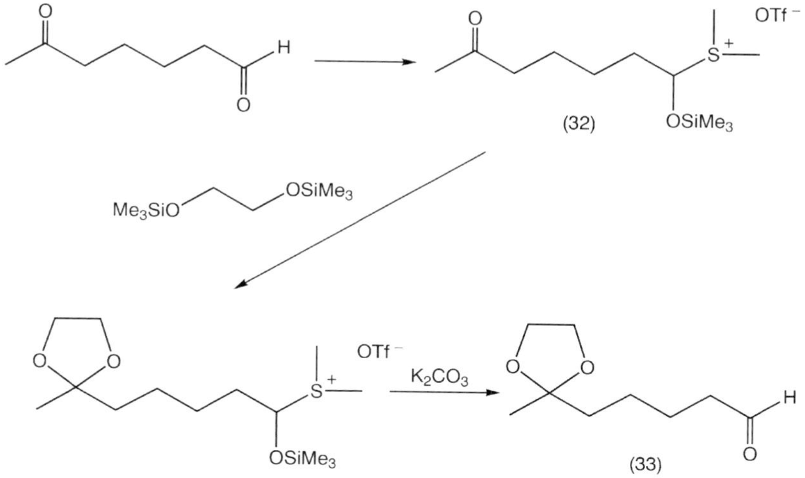

(32)

(33)

3.4 Thioacetals as protecting groups

Mercaptoethanol (34), with the more nucleophilic sulfur, produces 1,3-oxathiolanes (35) from aldehydes and ketones.[106] Although isomers

(34)

(35)

(36)

are possible, a single isomer is often formed. Alternative methods to acid-catalysed hydrolysis, based on sulfur chemistry, are also available. For example oxidation of the 1,3-oxathiolane (35) to the sulfone (36) permits the extrusion of sulfur dioxide in an electrocyclic process.

Thioacetals are formed from a thiol and an aldehyde or ketone in the presence of an acid catalyst. Unlike acetal formation, the equilibrium between a ketone and a thioacetal is normally significantly in favour of thioacetal formation. It is also possible to achieve the monothioacetalisation of polycarbonyl compounds using thiosilanes. The aldehyde of sugars can be protected in the open chain form as the dithioacetal, and thus ribose (37) forms the dithioacetal (38).[107]

(37)

(38)

Cyclic dithioacetals, (40) and (42), are formed from ethanedithiol (39) and propane-1,3-dithiol (41). The reagents are extremely unpleasant smelly compounds to handle. Although protecting an aldehyde against nucleophilic addition, the thioacetal renders what was formerly the aldehydic hydrogen acidic and the resultant carbanion achieves some stability from the presence of the adjacent sulfur atoms. This carbanion is a useful nucleophile and it may be, for example, alkylated. Thus the reactivity of the aldehyde carbon atom has been reversed from being sensitive to nucleophilic attack to behaving as a nucleophile. Just as with 1,3-dioxolanes, steric factors play a role in determining the selectivity of these reagents and it is possible to prepare partially protected derivatives such as (44) from (43), allowing the selective modification of the more hindered carbonyl group.[108]

Thioacetalisation using ethanedithiol and Lewis acid catalysis (boron trifluoride etherate) can bring about double-bond isomerisation. This can be avoided by using ethylenedithiobistrimethylsilane (45) and zinc iodide

as a catalyst.[109] The conversion of the sesquiterpenoid nootkatone (46) to valencene (47)[110] provides an example of this. The conventional system brought about double-bond isomerisation affording (48) whilst the use of

the thiosilane and zinc iodide left the double bond in place giving
(**49**). Unlike the dioxolane example (**17–18**), the double bond of the
unsaturated ketone did not migrate to C-5.

(46) R = O
(47) R = H$_2$

(48)

(49)

The unmasking of the carbonyl group from the thioacetal is less
satisfactorily achieved with acid compared with an ordinary acetal. Salts
of heavy metals such as mercuric acetate and cadmium carbonate which
are thiophilic are used under neutral conditions. A facile solid-state
cleavage to the parent ketone by using Montmorillonite K10 clay
supported iron(III) nitrate (clayfen) has been reported.[111] Alternatively,
the thioacetal may be reductively removed with Raney nickel in which
case the product is a methylene rather than the free carbonyl
compound.[112,113]

The cleavage of 1,3-dithianes may be facilitated by alkylating the sulfur
with methyl iodide.[114] A similar result may be achieved by using the
bromodimethylsulfonium ion. This may be generated in situ from t-butyl
bromide and dimethylsulfoxide.[115] The deprotection reaction may proceed as shown in Scheme 3.21 (**50–52**). The decomposition of a thioacetal by using dimethylsulfoxide and iodine[116] may proceed similarly.
However, the cleavage of 1,3-dithianes with antimony pentachloride may
occur by a series of single electron transfer steps.[117] 1,3-Dithiane
protecting groups may also be removed oxidatively with reagents such
as N-bromo- or N-chloro-succinimide, chloramine-T or ceric ammonium
nitrate.

3.5   Enol ethers and enamines as protecting groups

Enol ethers modify the reactivity of the carbonyl group but are
themselves too reactive to act as protecting groups except in special

(50) (51) (52)

circumstances.[118,119] Two situations where they are used are, first, when they form part of a diene, in which they represent a masked αβ-unsaturated ketone (53) and, second, when they are formed by the Birch reduction of an aromatic ring (54). Silyl enol ethers have been used in a

(53)

(54)

similar manner. The diene may participate in a Diels–Alder reaction and the masked ketone is subsequently released by treatment with acid. The diene, 1-methoxy-3-(trimethylsilyloxy)-1,3-butadiene (56) (Danishefsky's diene),[120] prepared by silylation of the methoxyenone (55), reacts with a range of dienophiles affording, after removal of the protecting group, carbocyclic and heterocyclic products (for example 57 and 58).

The use of an enol ether as a protecting group for a β-diketone (59) can alter the regiospecificity of their alkylation reactions, providing a general synthesis of alkylcyclohexenones (60).[121] For example the enol ether of dihydroorcinol (61) was successively alkylated with methyl iodide and isopentenyl bromide to give (62) in the course of a synthesis of the sesquiterpenoid nootkatone.[109]

The stability of enamines to Grignard conditions has been utilised in the protection of carbonyl groups in selected circumstances. For example Δ⁴-3-keto steroids (63) have been protected as the pyrrolidinyl enamines (64) allowing a distinction to be made between a cyclohexenone and a more hindered cyclopentanone in the preparation of (65). These derivatives

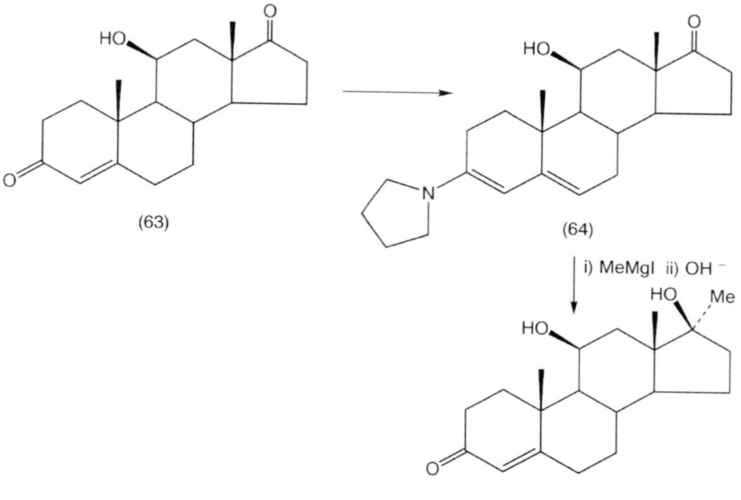

(63)

(64)

i) MeMgI ii) OH⁻

(65)

have been shown to be stable to acid when the iminium salts (66) are formed but they are rapidly cleaved by alkali.[122]

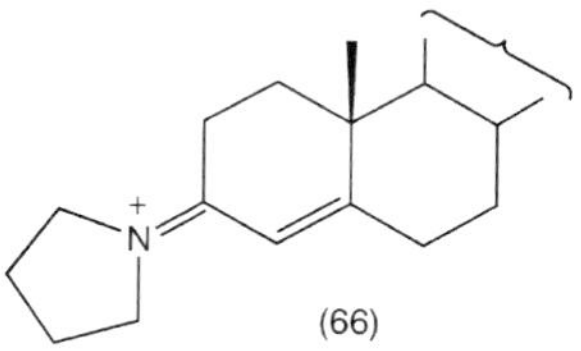

(66)

3.6 Nitrogen derivatives as protecting groups

The selective protection of carbonyl groups using nitrogen derivatives such as the oxime, the hydrazone and the semicarbazone has been explored but a difficulty with these derivatives lies in their cleavage. There are a number of examples of their use. In the course of a synthesis[123] of the sesquiterpenoid sativene, the alkene (67) was converted by hydroboration to the alcohol (69). In order to avoid reduction of the ketone, it was protected as its 2,4-dinitrophenylhydrazone (68). Ozonolysis sufficed to remove the protecting group after the hydroboration.

Trimethylsilyl chloride:sodium nitrite or t-butyl nitrite are useful reagents for regenerating carbonyl compounds from their oximes.[124]

(67) 2,4-DNPH → (68)

i) B₂H₆
ii) H₂O₂, NaOH

(69) O₃

The same reagents can be used in the hydrolysis of semicarbazones. Ceric ammonium nitrate or phenyliodoso-acetate[125] in aqueous acetonitrile will oxidatively cleave semicarbazones. Other oxidising agents which have been used for this include manganese triacetate, sodium hypochlorite[126] and ammonium persulfate supported on clay.[127] Another method is based on an exchange reaction with pyruvic acid.[128] The presence of the carboxyl group in the pyruvic acid means that the residue of the protecting agent can be separated from other neutral compounds. The regeneration of ketones from toluene-*p*-sulfonylhydrazones requires more vigorous conditions, although the Envirocat EP2G® heterogeneous acid catalyst is effective.[129]

An interesting feature of O-methyloximes of aromatic aldehydes (**70**) is that the electron-donating properties of the methoxyimine group result in

(70)

chlorination and bromination of the aromatic ring taking place at the *ortho* and *para* positions (see arrows below the structure).[130] In these examples the protecting group was subsequently removed by treatment with acid.

N,N-Dimethylhydrazones[131] may be used as protecting groups for ketones. Their cleavage is facilitated by conversion to the trimethyl-hydrazinium salt which is sensitive to mild alkaline hydrolysis. An alternative method of cleavage involves oxidation with peracid to the

N-oxide which is subject to mild alkaline hydrolysis. Hydrazones may be cleaved by hypervalent organoiodine reagents.[132]

A ketone can be protected as a cyanohydrin. The exothermicity associated with the formation of a silicon–oxygen bond can be used to enhance the formation of silyl derivatives of cyanohydrins from ketones which do not easily give cyanohydrins (**71–72**).[133] An example of their use involves the formation of an unstable dienone (**73–74**).[134]

(71)　　　　　　　　　　　　　　　　(72)

(73)　　　　　　　　　　　　　　　　(74)

3.7　The protection of 1,2- and 1,3-diketones

The selective protection of 1,2- and 1,3-dicarbonyl compounds can be achieved in various ways. For example ethyl acetoacetate (**75**) reacts with ethylene glycol to form a 1,3-dioxolane (**76**) allowing modification of the ester grouping.

(75)　　　　　　　　　　　　　　　　(76)

β-Formylketones (**77**), which exist predominantly as their tautomeric hydroxymethyleneketones (**78**), can be masked as their enol ethers (**79**) or enamines (**80**). A consequence of this is that a methylene adjacent to a carbonyl group can be blocked since the corresponding β-formylketones are readily prepared by condensation with ethyl formate. Hence the regiospecificity of alkylation adjacent to a ketone can be controlled. Thus the decalone (**81**) was alkylated via (**82**) at the angular position to give

OMe
MeI, K_2CO_3
O
H
OH
O
O
O
Me
N
Ph
PhNHMe
O

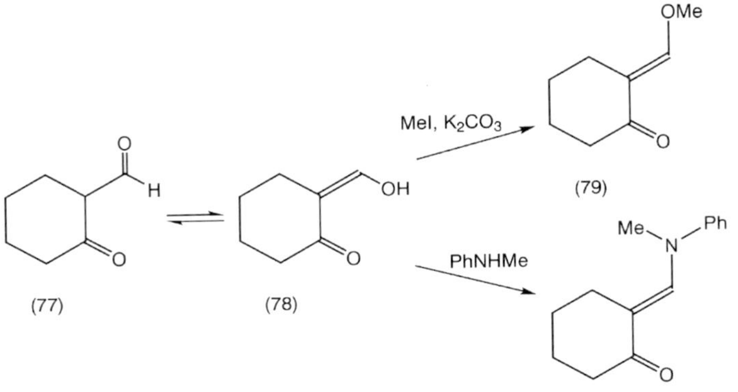

(77) (78) (79) (80)

(**83**). Similarly, 2,2-dimethylcyclohexanone (**87**), free of 2,6-dimethylcyclohexanone (**88**), was prepared from 2-methylcyclohexanone (**84**) by the sequence illustrated in (**84**)–(**87**).[135]

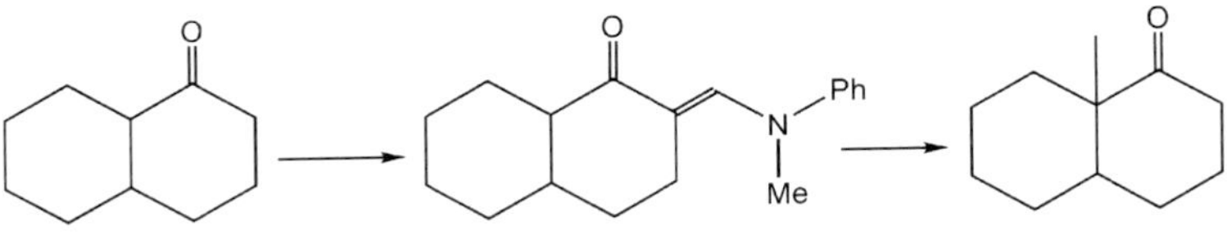

O
O
Ph
N
Me
O

(81) (82) (83)

i) HCO_2Et, NaOEt
ii) MeI, K_2CO_3
OMe
MeI, $NaNH_2$
OMe
O
O
O
H_3O^+
O
O

(84) (85) (86) (87) (88)

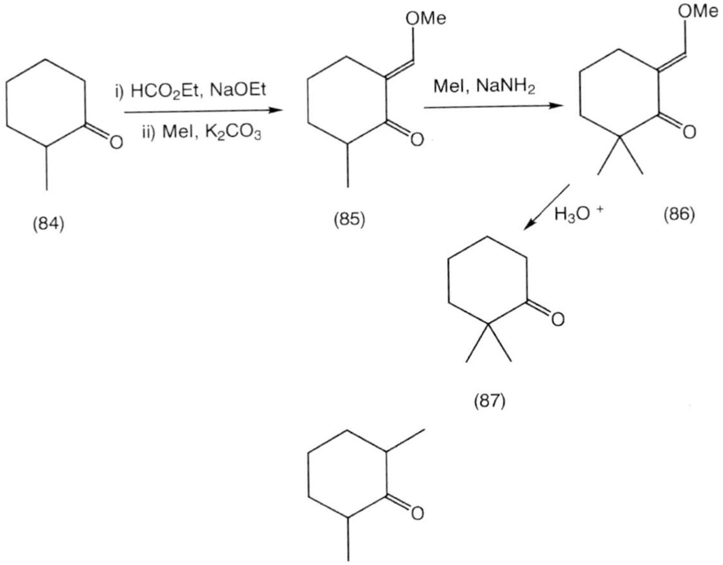

4 The protection of carboxylic acids

4.1 Introduction

Carboxylic acids react with many basic catalysts and reagents and consequently the acidic hydrogen needs to be masked. Carboxylic acids are normally protected as their esters, often the methyl esters. However, esters do not provide complete protection for all reaction conditions— for example methyl esters may be reduced by lithium aluminium hydride. The older methods of esterification involved vigorous conditions or toxic reagents and the hydrolysis of esters, particularly those of hindered carboxylic acids, required drastic reaction conditions. In this chapter we will consider some of the older methods of esterification showing how these have been improved by activating either the alcohol or the acid. Methods of hydrolysis and modifications to the structure of the esters to facilitate their removal will be discussed.[136] The importance of carboxyl protection in peptide synthesis has led to the development of a number of reagents, some of which have found general application.

4.2 Methods of esterification

The acid catalysed esterification of carboxylic acids is an equilibrium process (**1–2**) in which water is produced. In order to improve the yield of

the ester, methods have been devised to remove the water as it is formed. These include the azeotropic distillation of the water. Some acid catalysts, such as sulfuric acid, also act as dehydrating agents. These conditions, however, can be drastic and may lead to reactions at other centres in the molecules including dehydration of the esterifying alcohol. The acid-catalysed cleavage of a number of acetals and orthoesters can afford reactive species which facilitate the formation of esters. Thus acetone dimethylacetal (3) brings about the esterification of acids in the presence of a catalyst. Trimethyl and triethyl orthoacetate (4)[137] and N,N-dimethylformamide dimethylacetal (5) function as esterifying agents in a similar way.[138] The lone pairs on the oxygen or nitrogen atoms play an important role in the generation of the reactive intermediates.

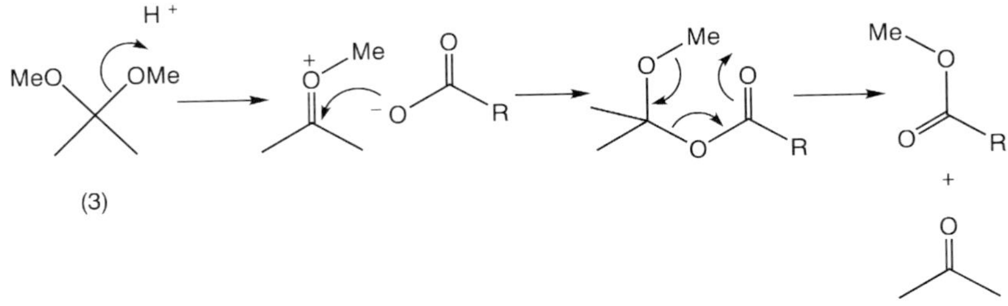

MeC(OEt)₃

(4)

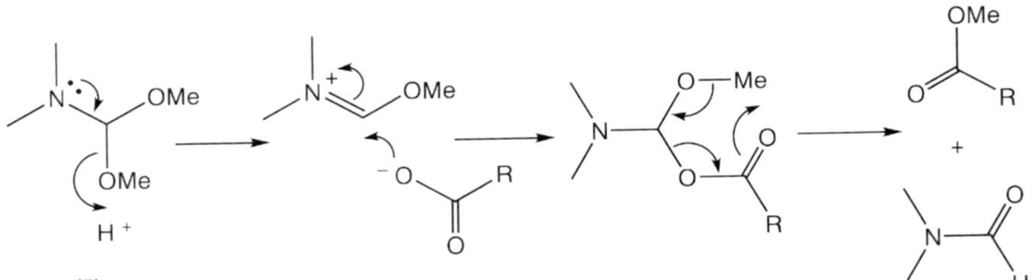

Diazomethane (7), while of value as an esterifying agent, is both toxic and explosive, precluding its use on a large scale. On the small scale it is a very effective methylating agent. Diazomethane is prepared by treating Diazald® (N-methyl-N-nitrosotoluene-p-sulfonamide) (6) with potassium hydroxide and it is commonly used in solution in ether.[139] Other diazomethane precursors include N-nitroso-N-methylurea (8) and 1-methyl-3-nitro-1-nitrosoguanidine (MNNG) (9). The structure of diazomethane can be described in various resonance forms, including (7a), (7b) and (7c). Diazomethane reacts with carboxylic acids as in (10). A safer alternative to diazomethane is trimethylsilyldiazomethane (11).

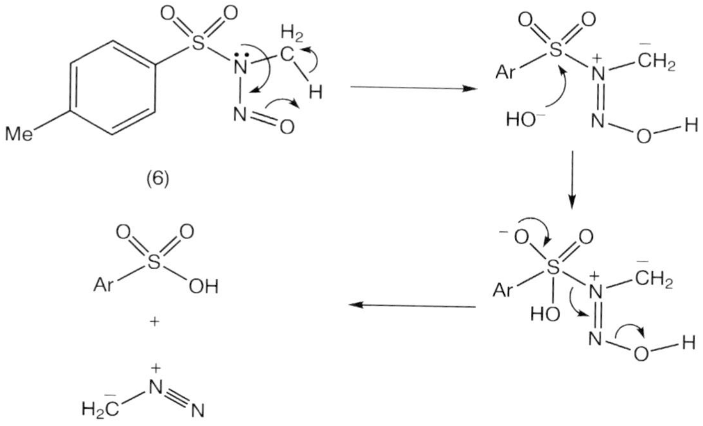

(6)

(7)

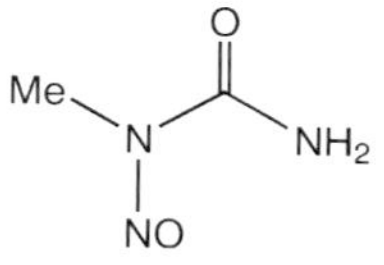

(8)

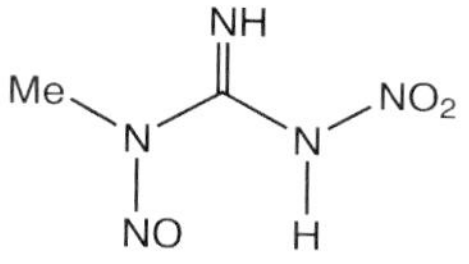

(9)

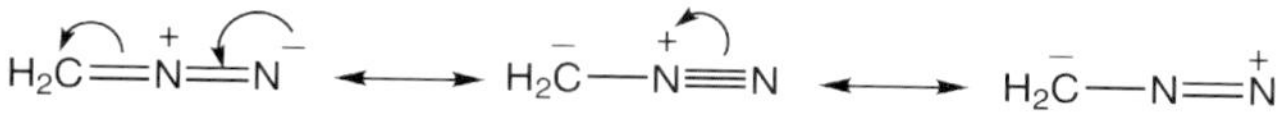

(7a)   (7b)   (7c)

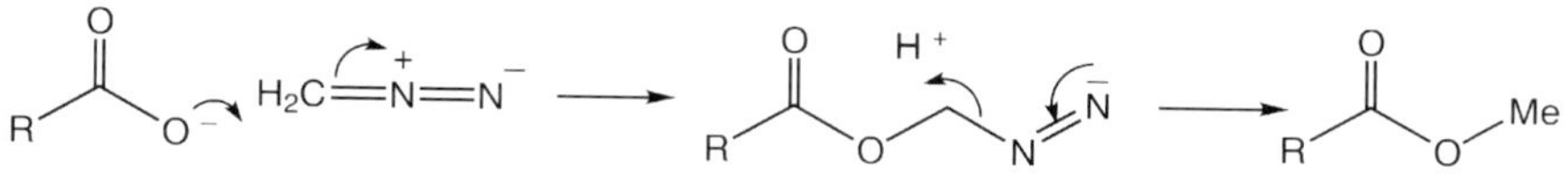

(10)

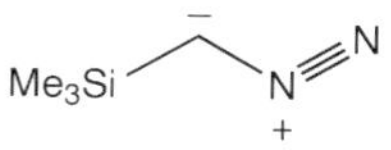

(11)

4.3 Activation of the esterifying alcohol

Activation of the alcohol by formation of the trichloroimidate (12) by using trichloroacetonitrile may facilitate the esterification by converting the alcohol into a better leaving group. This reaction leads to inversion of configuration of the alcohol. A similar result is obtained when triphenylphosphine (13) and ethyl azodicarboxylate (14) are used in the Mitsunobu esterification reaction.[140] The initial adduct (15) activates the phosphorus. The driving force is the formation of an alkoxyphosphonium salt (16) which undergoes a nucleophilic substitution by the carboxylate anion, again with inversion of configuration.

Alcohols may be activated by conversion to the corresponding alkyl halide. The reaction of alkyl halides with the silver salt of the carboxylic acid can also be used to prepare esters. A modification of this method involves treating the acid with caesium fluoride and methyl iodide.[141] The nucleophilicity of the carboxylate may be enhanced by reducing the extent of close ion-pair formation by replacing the metal salt with a quaternary alkyl ammonium salt. Alternative methods involve using a dipolar aprotic solvent such as dimethylsulfoxide or dimethylformamide, a crown ether or a phase-transfer catalyst.

A benzyl ester is a useful protecting group for a carboxylic acid since it may be removed by mild hydrogenolytic conditions. Benzyl esters are prepared from the silver salts and benzyl bromide or by methods based on activating the alcohol via benzyl trichloroacetimidate.[142] A mild method[143] for preparing benzyl esters involves activation of the alcohol through its O-benzyl-S-propargyl xanthate (17). The activation is based on the isomerisation of propargylxanthates to allenes (18) and the acid-catalysed cyclisation of these to reactive intermediates such as (19).

4.4 Activation of the carboxylic acid

The lone pair of the hydroxyl oxygen diminishes the susceptibility of the carbonyl group of an acid to nucleophilic addition. Hence acid anhydrides or mixed anhydrides provide a useful method of generating esters. Interaction of the second carboxylate with the lone pair (20) makes it less readily available to conjugate with the original carbonyl group. One variation of this method is to prepare a mixed anhydride with a

relatively strong and hindered aromatic carboxylic acid. A mixed anhydride with 2,4,6-trichlorobenzoic acid is prepared by reacting the carboxylic acid with 2,4,6-trichlorobenzoyl chloride (21) in the presence of dimethylaminopyridine and triethylamine.[144] This mixed anhydride undergoes alcoholysis at the nonaromatic carbonyl group to form the ester.

A number of reagents activate the carboxyl group to nucleophilic attack by the alcohol by increasing the electron deficiency of the carbonyl carbon. The electron-withdrawing properties of these substituents may also facilitate the collapse of the tetrahedral intermediate (22) and favour

(22)

ester formation. The structure of these reagents must facilitate coupling of the nucleophilic carboxylate with the activating agent but the carboxyl carbon must also remain the most sensitive part of the molecule to attack by the nucleophilic oxygen of the alcohol. This is arranged by making the activating group 'softer' in character. Finally, solubility may be used to favour the separation of the residual part of the activating group once it has performed its task.

One of the best known activating reagents is dicyclohexylcarbodiimide (DCC) (23) in which the central diimide carbon is electron deficient and susceptible to nucleophilic addition by the carboxylate. The adduct (24) behaves as a mixed anhydride but since nitrogen is less electronegative than oxygen the imino-carbon atom is less readily attacked by the alcohol. The collapse of the tetrahedral intermediate (26) is facilitated by the formation of a urea (25). This particular urea is relatively insoluble

and is often easily separated from the ester. Esterification by this method can be catalysed by the addition of a small amount of dimethylamino-pyridine.[145] Separation problems can be overcome by using water-soluble and acid-soluble carbodiimides[146,147] as exemplified by 1-ethyl-3(3-dimethylaminopropyl)carbodiimide (27) where salts of the dimethyl-aminopropyl moiety are water soluble.

A number of heterocyclic compounds have been used to activate carboxylic acids. Since the bromine of a 2-bromopyridinium salt (28) is readily displaced, it behaves like an imino or an acyl halide and the pyridine ring is easily coupled to a carboxylate to give (29). The formation of a pyridone (31) provides the driving force for the collapse of the tetrahedral intermediate (30) in the formation of the ester.[148] O,O'-Di(2-pyridyl)thiocarbonate (DPTC) (32) is an efficient reagent[149] for the

preparation of the activated 2-pyridyl ester intermediates (34). The thionocarbonate not only enhances the electron deficiency at C-2 of the pyridine ring but also facilitates the collapse of the tetrahedral intermediate (33) by the expulsion of COS and a 2-pyridinone (35).

Appel's salt (**36**), prepared from sulfur dichloride and trichloroaceto-nitrile, activates carboxyl groups by forming the derivative (**37**).[150] Esterification involves the formation of (**38**).

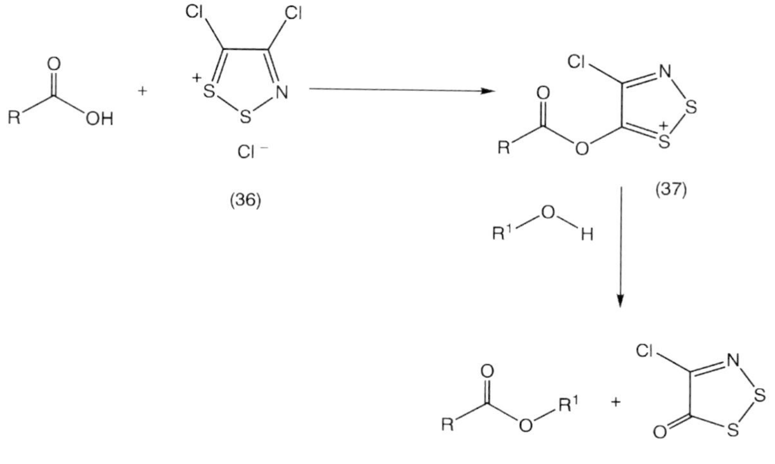

(36)   (37)

(38)

Because of the strength of the silicon–oxygen bond, trimethylsilyl derivatives (**39**) make a useful way of activating a carboxyl group.[151,152] Methyl esters can be prepared by treating the carboxylic acid with trimethylsilyl chloride in methanol.

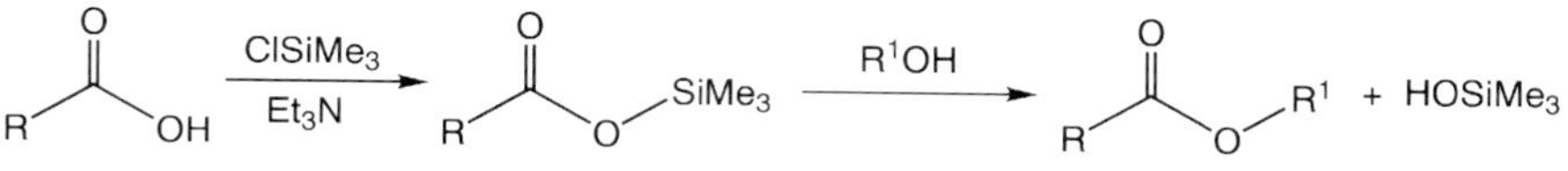

(39)

Acid chlorides react with alcohols to form esters. Many of these reactions are catalysed by pyridine or, better, dimethylaminopyridine (**40**).[153] This functions through the formation of an acylpyridinium salt (**41**) which is highly susceptible to nucleophilic attack.

4.5   The hydrolysis of esters

There are two underlying general mechanisms for ester hydrolysis involving either alkyl or acyl oxygen fission. One involves attack on the alkyl carbon of the ester with a nucleophilic displacement of the carboxylate (**42**) and the other involves addition to the carbonyl group to

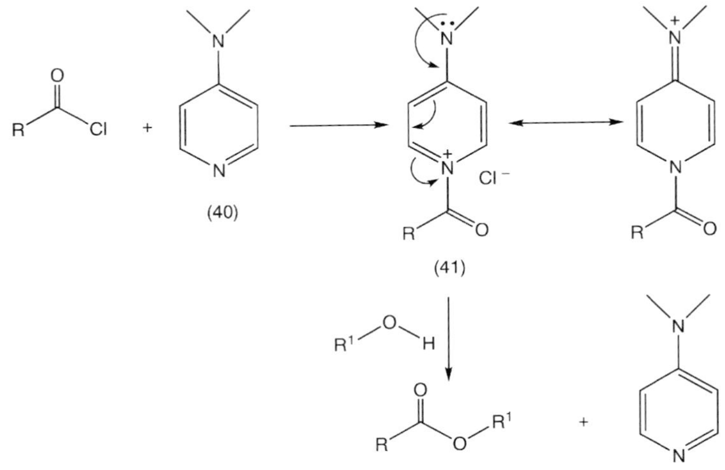

(40)

(41)

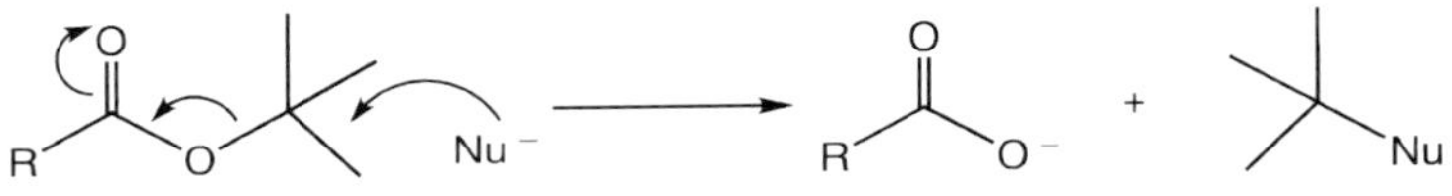

(42)

form a tetrahedral intermediate (**43**) which collapses with the expulsion of the alcohol. The efficacy of these methods depends very much on stereo-electronic features involving not only the trajectory of the incoming

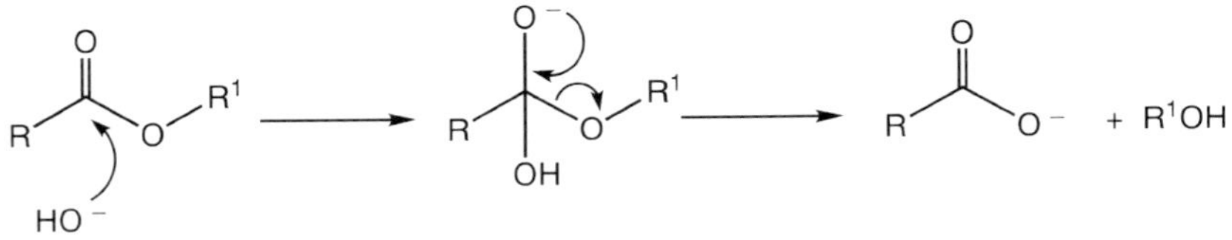

(43)

nucleophile but also the bulk of the tetrahedral intermediate. Various methods have been developed for enhancing the cleavage of esters based on these two pathways.[154]

The S_N2 displacement of the carboxylate may be favoured by using anhydrous lithium iodide in collidine or lithium propanethiolate (**44**). The lithium coordinates to the ester oxygen enhancing its leaving-group character, whilst the soft nucleophile attacks the alkyl carbon of the ester.[155] A solvent of relatively low dielectric constant is used so that the lithium will coordinate with the substrate rather than with the solvent. Dry lithium iodide is used to avoid competitive binding to water.

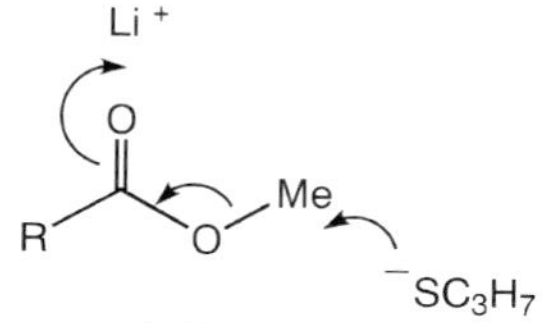

(44)

The reactivity of the carboxylate is increased by coordination to a Lewis acid such as aluminium trichloride and the displacement is then carried out with a powerful nucleophile such as ethanethiol (45).[156]

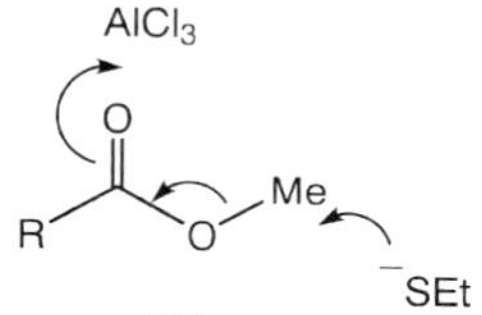

(45)

Trimethylsilyliodide is a useful reagent for effecting ester cleavage.[157] Its action is rationalised in terms of the theory of hard and soft acids and bases. Silicon is regarded as hard and reacts with the oxygen of the ester whereas iodine, which is soft, reacts with the carbon of the alkyl group, of, for example, (46). Trimethylsilyl iodide can be prepared in situ from trimethylsilyl chloride and sodium iodide. Boron tribromide can also be used as a catalyst for ester hydrolysis.[158]

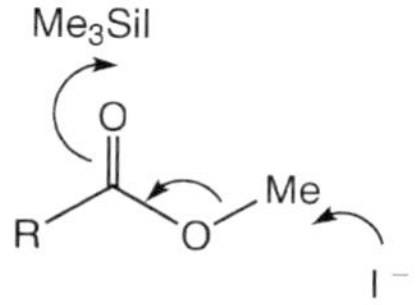

(46)

The efficacy of various methods of hydrolysis of esters can be assessed by the ease with which two hindered esters, methyl mesitoate (47)

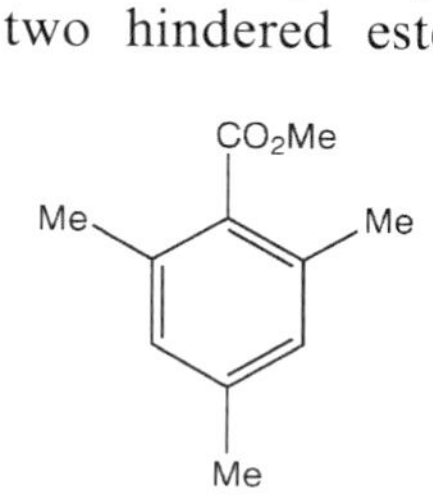

(47)

and methyl O-methylpodocarpate (**49**) are hydrolysed.[159] Methyl
O-methylpodocarpate possesses an axial carboxyl group which is
hindered by a 1:3-diaxial interaction with the angular methyl group. It
is sufficiently hindered that the ester is prepared by using sodium
hydroxide and dimethylsulfate (**48**). The carboxylate anion attacks the

(48)   (49)

dimethylsulfate as a nucleophile but the carbonyl group of the resultant
ester is sufficiently hindered for the tetrahedral intermediate involved in
its hydrolysis not to be formed. Hence the ester formation proceeds to
completion and is not accompanied by the reverse hydrolysis to the acid.
A number of ester groups have additional functionality to facilitate
their cleavage. For example an aromatic ring stabilises an adjacent
carbonium ion thus facilitating the acid-catalysed or hydrogenolytic
cleavage of benzyl esters. In some examples p-methoxybenzyl (**50**) or (4-
methoxyphenyl)ethyl esters (**52**) have been used in which the electron-
donating p-methoxy group increases the ease of cleavage (see **51**).[160] The
diphenylmethyl protecting group derived from the reaction of benzhydrol

(50)   (51)

(52)   (53)

and the carboxylic acid is an extension of this principle.[161] The sensitivity of the benzylic position to radical attack has led to N-bromosuccinimide being used as a reagent for cleaving benzyl esters.[162]

The tertiary butyl ester (53) is an important protecting group for carboxylic acids because of the hindrance it affords to nucleophilic attack on the carbonyl group and its sensitivity to cleavage under mild acid-catalysed conditions. This acid lability means that t-butyl esters can be removed in the presence of less reactive esters such as the methyl ester. However, t-butyl esters are more difficult to prepare. They can be prepared by the reaction of the carboxylic acid chloride with t-butanol in pyridine or with an alkali metal t-butoxide. Reaction with isobutylene and an acid catalyst involves working under pressure. An alternative to this involves using a catalyst of concentrated sulfuric acid dispersed on magnesium sulfate in t-butanol.[163] The carboxylic acid may also be esterified directly with a t-butyl halide, dimethylformamide di-t-butyl acetal or by activation with DCC or carbonyldiimidazole followed by reaction with t-butanol.

A cyanomethyl ester (54) may be prepared from the acid and chloroacetonitrile in the presence of triethylamine. Neighbouring-group participation by the less hindered nitrile as in (55) may facilitate the hydrolysis of the ester.

The β-(trimethylsilyl)ethoxymethyl ester (56) is a useful protecting group for carboxylic acids whenever mild nonhydrolytic deprotection is required. The protecting group is removed with fluoride ion or with magnesium bromide etherate.[164] The latter is compatible with the presence of other protecting groups such as the carbamates (Boc, Cbz and Fmoc) as well as fluoride-ion sensitive silyl ethers.

A 2-cyanoethyl ester is another protecting group that can be removed by a fragmentation reaction.[165] This protecting group is introduced by treating the acid with ethylene cyanohydrin in the presence of DCC and dimethylaminopyridine as a catalyst. The cleavage of the ester is achieved under mild conditions by using tetrabutylammonium fluoride in dimethylformamide. The trifluoroethyl group has been used as an esterifying group to protect carbohydrate sulfates. It was introduced using 2,2,2-trifluorodiazoethane and removed with potassium t-butoxide in t-butanol.[166] Neopentyl esters have been used as protecting groups for arylsulfonic acids.[167]

Phenacyl esters (57) and β-trichloroethyl esters (58) may be reductively cleaved with zinc in an elimination reaction. Phenacyl esters may also be

(57)

(58)

cleaved photochemically.[12] An example of the use of the phenacyl protecting group is in the preparation of gibberellin plant hormones (59)[168] which are unstable in acidic and alkaline conditions. Acetol

(59)

($CH_3CO \cdot CH_2OH$) is a variant on the phenacyl group which can be cleaved in some instances with tetrabutylammonium fluoride.[169]

Fragmentation reactions have played a part in the removal of other ester groupings, exemplified by the cleavage of β-trimethylsilylethyl esters (60). This group has been used to protect not only carboxylic acids but also phosphates in oligonucleotide synthesis.

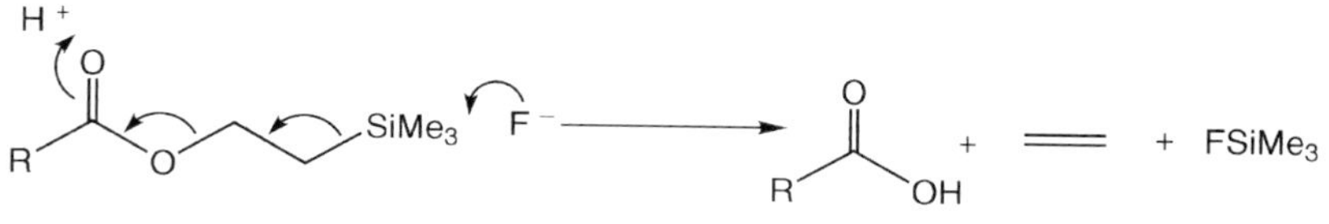

(60)

Allyl esters (61) can be cleaved by organometallic reagents. The use of allyl malonates provide an example of this.[170] The decarboxylation of

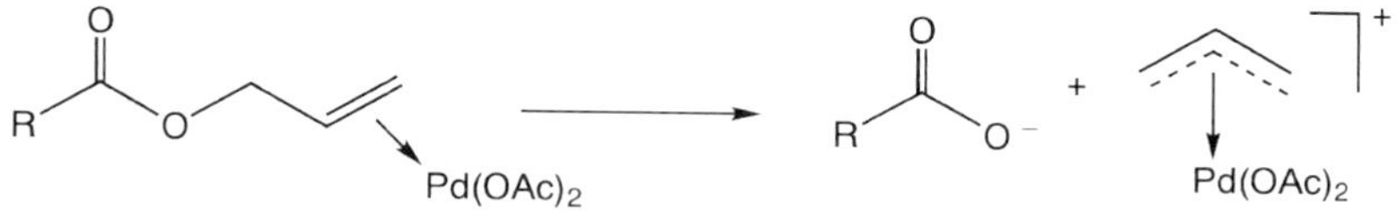

(61)

malonic ester derivatives is an important method for the preparation of substituted carboxylic acids but typically requires harsh conditions. A method based on palladium-catalysed deallyloxycarbonylation in the presence of ammonium formate provides a milder alternative. This is exemplified[171] in the preparation of the β-lactam (62–63).

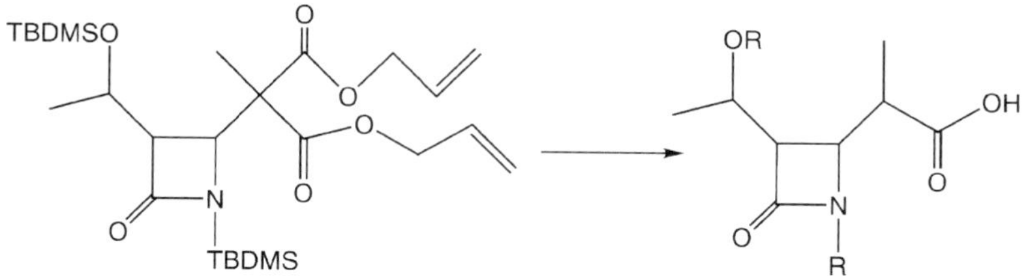

(62)   (63)

Members of another group of unsaturated esters that are readily cleaved are 3-methylbut-2-enol esters (64).[172] This protecting group can be introduced by reacting the acid with the unsaturated alcohol and DCC in the presence of dimethylaminopyridine. Iodine makes a simple deprotecting agent. A mechanism for the deprotection is shown in (65).

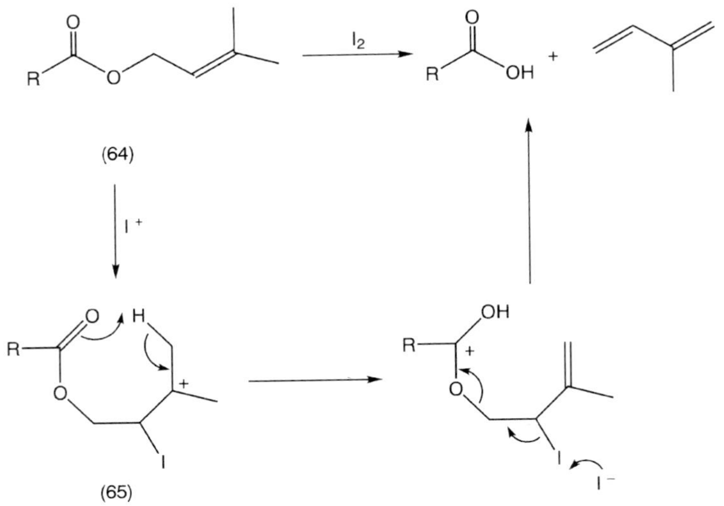

(64)

(65)

A method which has rarely been exploited involves the use of photolabile protecting groups exemplified by the *o*-nitrobenzhydryl esters (**66**). Photochemical rearrangement gives the derivative (**67**), which rapidly collapses to the carboxylic acid and benzophenone.[12]

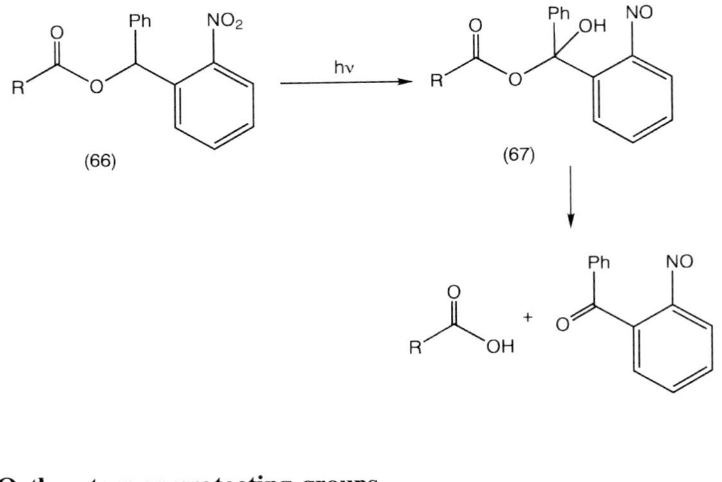

(66)   (67)

4.6   Orthoesters as protecting groups

Esters, however bulky the alkyl group, do not provide complete protection against nucleophilic attack at the carbonyl group. The conversion of carboxylic acids to orthoesters provides protection of the carbonyl group against, for example, reduction by lithium aluminium hydride.[173] The trioxoadamantyl protecting group in (68) has been used for the preparation of labelled mevalonic acids (69).[174] The orthoester protecting groups are removed by acid-catalysed hydrolysis.

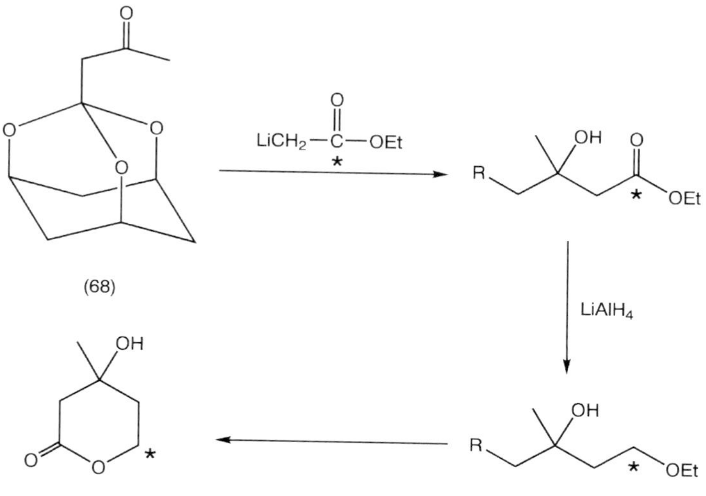

(68)

(69)

5   The protection of the amino group

5.1   Introduction

There are a number of chemical properties of the amino group which may need masking during a synthesis. The amino group is both a powerful nucleophile and a base. An ammonium salt formed by protonation of an amine has different chemical reactions, for example directing effects in aromatic substitution, compared with the parent amine. The hydrogen atoms of a primary or secondary amine are sufficiently 'active' to react with a Grignard reagent, and the nitrogen lone pair may be involved in complexing with a metal ion, including copper, thereby affecting the reactions of organometallic compounds such as organocuprates. Amines are also susceptible to oxidation which can lead, for aromatic compounds, to decomposition. In the synthesis of peptides it is necessary to mask the amino group of one amino acid to allow the nitrogen of a second amino acid to participate in peptide bond formation. The masking group must be removed without disturbing the existing peptide (amide) bonds. In other syntheses, for example with amino sugars, a differentiation has to be made between the reactivity of a hydroxyl and an amino group.

The conversion of an amine (1) to an amide (2) diminishes the nucleophilicity of the nitrogen. However, the subsequent cleavage of the

$$RNH_2 \quad \longrightarrow \quad$$

(1)   (2)

amide of a protected amine in the presence of peptide bonds, has led to the development of a special range of protecting groups. Many have found general application outside peptide synthesis. In this chapter we will show how alkylation and acylation can be used to protect an amine.

5.2   Alkyl and related protecting groups

When alkyl groups are added to an amine affording a tertiary amine, they protect the nitrogen against acylation by removing the N–H group.

However, it is important to introduce an alkyl group which can be selectively removed. A benzyl group (3) is easily introduced by using a

(3)   (4)

reactive benzyl halide and easily removed by hydrogenolysis. The bulky triphenylmethyl group (4) has been used for the same purpose. The N-silyl groups, for example N–SiMe$_3$ and N–SiMe$_2{}^t$Bu, can also protect an N–H. The free amine may be regenerated by treatment with tetra-butylammonium fluoride or other desilylating agents. N-Silyl derivatives are, however, more easily cleaved than the analogous O-silyl compounds and may even be used as silyl donors.[175]

Primary amines form cyclic disilazane adducts (stabase adducts) on reaction with 1,4-dichloro-1,1,4,4-tetramethyldisilylethylene (5) in the presence of a base.[176] An example of their use is in the preparation of alkylated amino acids from the ethyl ester of glycine (6). Thus 2-amino-4-pentynoic acid (8), which is an antimetabolite of L-methionine and leucine, has been prepared by the sequence (5)–(8). A variant for

substituted aromatic amines used an exchange reaction with 1,1,4,4-tetramethyl-1,4-bis(N,N-dimethylamino)-disilylethylene (9) catalysed by

(9)   (10)

zinc iodide to prepare the stabase adduct (10).[177] A benzostabase version (11) has also been introduced[178] to protect aromatic amines.

(11)

5.3   Imines and enamines as protecting groups

An imine (12), prepared by reacting an amine with an aldehyde or ketone, can also function as a protecting group. These are generally stable to

(12)

alkali but are sensitive to hydrolysis by acid. One example of the use of such a protecting group is in the synthesis of some 3-amino-4(5H)-pteridinones[179] where the terminal amine of 2-aminopyrazine-3-carboxyhydrazide (13) was blocked as its benzylidene derivatives (14). Cyclisation to the pteridinone (16) was effected by condensation with ethyl orthoformate in acetic anhydride. The protecting group was removed by brief treatment with cold dilute hydrochloric acid to give (15).

A primary amine can be protected against strong basic conditions by incorporating the nitrogen into a pyrrole ring. Thus a 2,5-dimethylpyrrole (18) can be prepared by condensation with hexane-2,5-dione (17) and the parent amine (19) regenerated at the end of a sequence by reaction with hydroxylamine hydrochloride.[180] The pyrrole (21) was used to prevent[181] an aniline (20) from coordinating to copper in the course of a copper(I)-mediated methoxylation (see 22) of the aryl iodide.

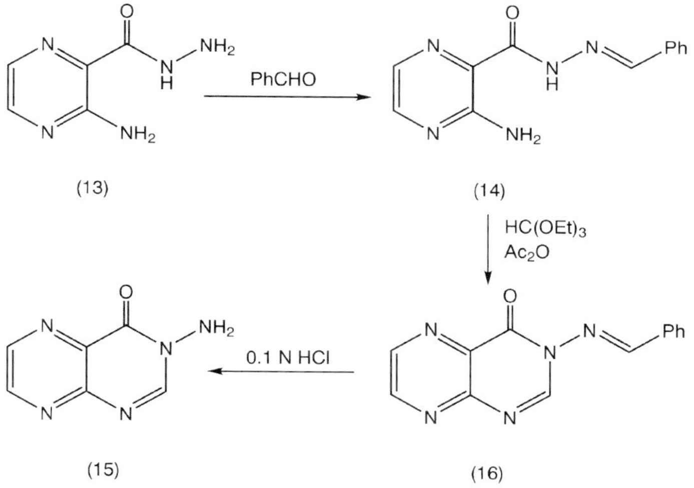

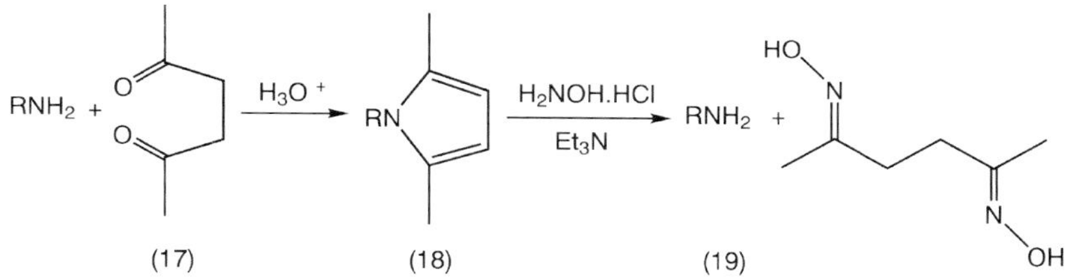

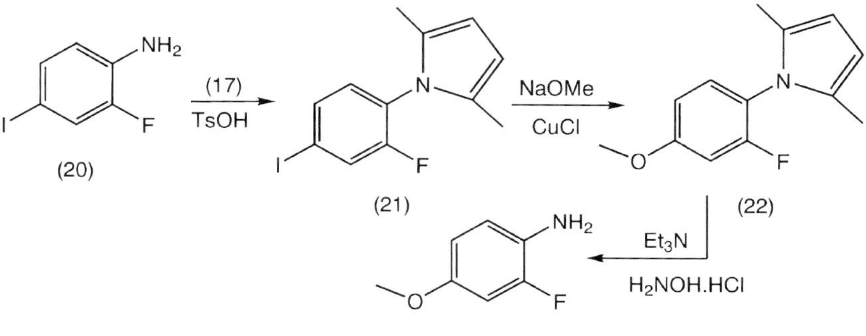

(13) (14)

(15) (16)

(17) (18) (19)

(20) (21) (22)

Bis[(triisopropylsilyl)oxy]pyrrole (BIPSOP) derivatives (23) are the silyl enol ethers of succinimides[182] and are stable to a strong base such as an organolithium or alkoxide. Deprotection involves two steps. Hydrolysis of the silyl enol ether occurs with dilute hydrochloric acid to reform the succinimide (24), and hydrolysis of this is effected with hydrazine hydrate.

A tertiary amine can be protected from quaternisation by converting the amine to the N-oxide. For example the nitrogen atom of morphine

(23)   (24)

(25) was protected as the N-oxide (27) and the phenolic hydroxyl group methylated to give codeine N-oxide (28). Codeine (26) was obtained by reduction of the N-oxide. This strategy has been applied to the synthesis of several alkaloids.[183]

(25) R = H   R = Me   (27) R = H

(26) R = Me   (28) R = Me

5.4   Amides as protecting groups for amines

The protection of amino groups as amides reduces the nucleophilic character of the nitrogen atom. The simplest protecting groups of this type are the N-formyl derivatives (29). These can be prepared by using formic acid:acetic anhydride mixtures.[184] The formyl group can be removed by acidic hydrolysis or by oxidation with hydrogen peroxide to give the labile carbamic acid (30).

(29)   (30)

Formamidomalonic acid diethyl ester (32) is used in the synthesis of amino acids. The protected amino group is introduced directly by the reduction of the nitrosation product (31) of diethyl malonate in formic acid. N,N-Dibenzylformamidines (34)[185] can be prepared from primary amines and N,N-dibenzylformamide dimethyl acetal (33). These protecting groups may be removed by hydrogenolysis.

(31)   (32)

(33)   (34)

The acetyl group is stable to oxidising agents and is often used to protect aromatic amines in nitration sequences as in the simple examples (35)–(38). The group may modify the chemistry so that mono- rather than polysubstitution occurs as, for example, in bromination. The parent amine is formed by acidic hydrolysis.

(35)   (36)   (37)   (38)

There are a variety of conditions for acetylating aromatic amines. Reactive aromatic amines may be acetylated with acetic anhydride in water containing a catalytic amount of a sulfuric acid. Alternatively, acetic anhydride in pyridine may be used. When a reactive amine is heated under reflux with acetic anhydride the diacetate may sometimes be formed. On the other hand more vigorous conditions may be required for the acetylation of deactivated aromatic amines such as nitroanilines.[186]

Sometimes it is necessary, as in the amino sugars, to regenerate selectively an amino group, leaving an ester, acetal or glycoside untouched. Selective hydrolysis may be achieved by trapping the tautomeric imidate form of the amide with triethyloxonium fluoroborate.[187] Thus the O-ethyl-acetamidium fluoroborate (40) obtained from 2-acetamido-1,3,4,6-tetra-O-acetyl-2-deoxy-α-D-glucopyranose (39), was sensitive to hydrolysis with mild aqueous sodium hydrogen carbonate to give the free amine (41). Enzymatic methods can also be used selectively to liberate an amine function

in the presence of an ester. For example penicillin G acylase will hydrolyse the phenylacetamido group from a nucleoside as in the conversion of (42) to (43).[188,189]

Deprotection of mono-, di-, and trichloro- or trifluoro-acetates utilises milder conditions than those required to cleave unsubstituted acetates.[190] The phenol glucoside of tyrosine (46) was prepared by coupling aceto-bromoglucose (45) with N-trifluoroacetyl-L-tyrosine ethyl ester (44) to give (47). The protecting groups in (47) including the N-trifluoroacetyl

group were removed by hydrolysis with barium hydroxide. Trifluoro-acetic anhydride, which is extremely reactive, is widely used for intro-ducing the trifluoroacetyl group. However, ethyl trifluoroacetate is a milder trifluoroacetyl transfer reagent which can be used[191] to differ-entiate between primary and secondary amino groups and even on steric grounds to acylate selectively one of several primary amino groups.

Neighbouring group participation can facilitate the hydrolysis of amides. An example is the use of a quinone redox system (48–50) in which an internal nucleophile (49) is generated in the reduction step.[192]

A phthalimide (51) is a useful group for protecting primary amines and indeed it forms the basis of the Gabriel synthesis of primary amines. It may be introduced by using phthalic anhydride or via the acid chloride of the hemimethyl ester of phthalic acid. A milder method utilises N-ethoxycarbonylphthalimide (54) in which the electron-withdrawing N-ethoxycarbonyl group renders the carbonyl groups of the phthalimide

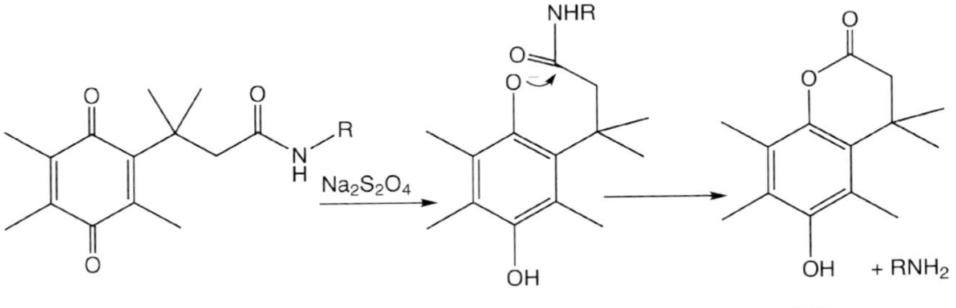

(48)   (49)   (50)

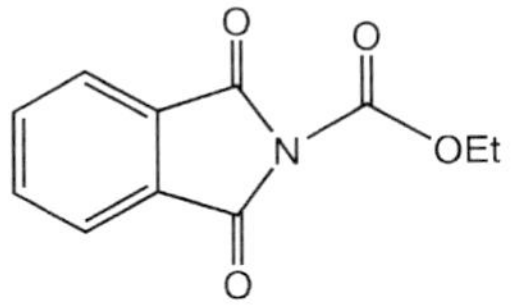

(54)

more sensitive to nucleophilic attack by the amine in the protection step.[193] The phthalimide grouping is stable to acid but it is hydrolysed by alkali. Hydrazine is particularly effective for this hydrolysis because the cleavage of the first amide by the formation of a hydrazide brings the second basic NH_2 into juxtaposition for hydrolysis of the second amide bond (52–53).[194] The reactivity of the imide to hydrolysis may be enhanced by using the 4-nitrophthaloyl group (55). The hydrolysis then proceeds with methylhydrazine in dimethylformamide at room temperature.[195]

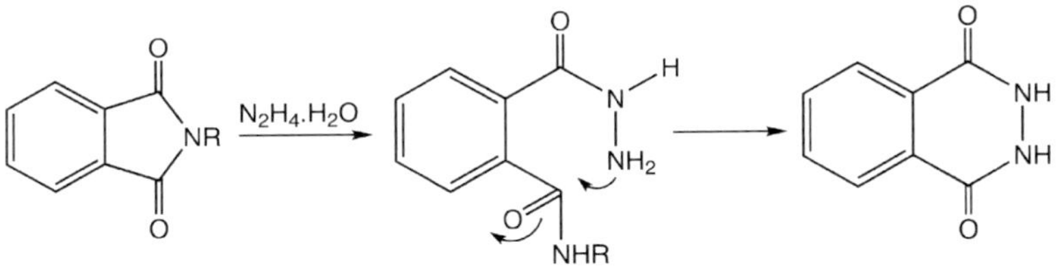

(51)   (52)   + RNH₂   (53)

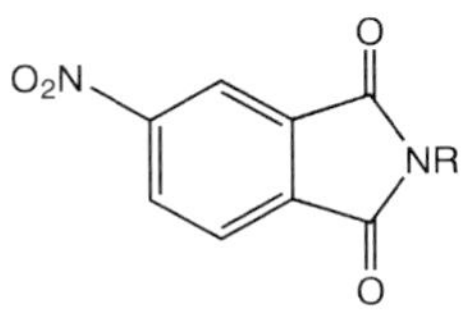

(55)

The toluene-*p*-sulfonyl group (**56**), which is introduced by using

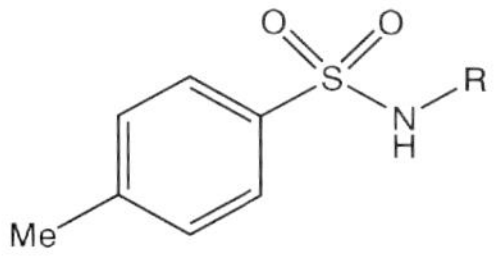

(56)

toluene-*p*-sulfonyl chloride, is a stable amine protecting group. The powerful electron-withdrawing effect of the arylsulfonyl group makes this protecting group particularly stable to acidic conditions. It can be removed reductively with sodium in liquid ammonia.[196] Samarium iodide in tetrahydrofuran is a one-electron reducing agent which has been applied not only to the removal of benzenesulfonamides but also to the cleavage of pyridine-2-sulfonamides.[197,198] A milder reagent which has been used for cleaving arylsulfonates is N-hydroxybenzotriazole (**57**).[199] This reagent behaves as a nucleophile (**58**). The repulsion of the lone pairs

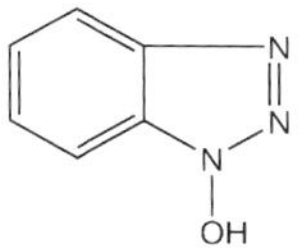

(57)

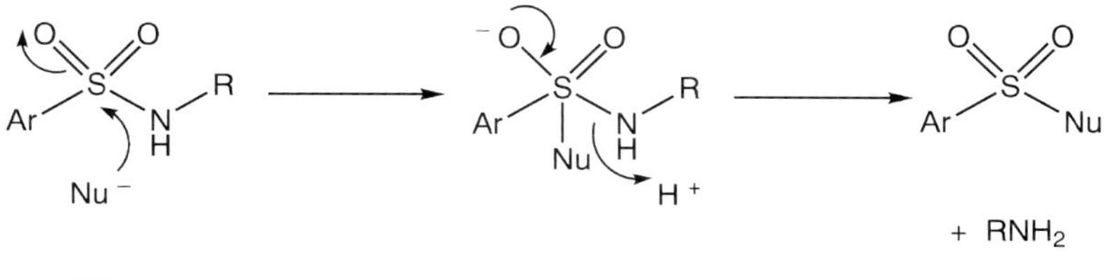

(58)

on the oxygen by the electron-rich heterocyclic ring increases the nucleophilicity of the oxygen. Both the toluene-*p*- and the *p*-methoxy-benzene-sulfonyl groups have been used to protect the imidazole N–H of histidine as exemplified by the derivative (**59**).

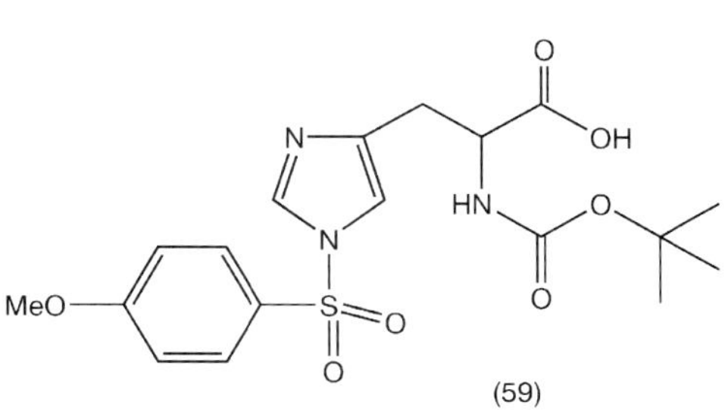

(59)

An alternative principle for diminishing the basicity of nitrogen utilises the electron-withdrawing o-nitrophenylsulfenyl group (60).[200] Deprotection relies on the ready formation of sulfur–sulfur bonds in the attack of a thiol to form a dithio derivative (61). Deprotection may also be brought about by acidic hydrolysis.

(60)   (61)

5.5   The protection of amino groups in amino acids[201]

There are a number of problems of protection associated with the synthesis of peptides. First, it is necessary to protect the amino group of one reactant (62) and the carboxyl group (63) of the other so that coupling is regiochemically specific. After the peptide bond has been

protect   activate   protect
      (62)   (63)   (64)

formed, the protecting groups must be removed from (64) under conditions which do not hydrolyse the peptide bond. Furthermore, it is important to have conditions which do not racemise the constituent amino acids. Racemisation (65–67) can occur via oxazolone formation (66). The enolic form (68) of the oxazolone has an aromatic character, thus favouring its formation. Finally, many of the amino acids have reactive functional groups in their side chain which require protection. These protecting groups often need to remain in place during several stages of a peptide synthesis and it was in this context that the concept of orthogonal protecting groups was introduced.[7]

A commonly employed protecting group of the amino-acid nitrogen is a carbamate (or urethane). A carbamate (69) combines the low nucleophilic reactivity of an amide with facile decarboxylation (70–71) in the deprotection step. The simplest methoxy and ethoxycarbamates are

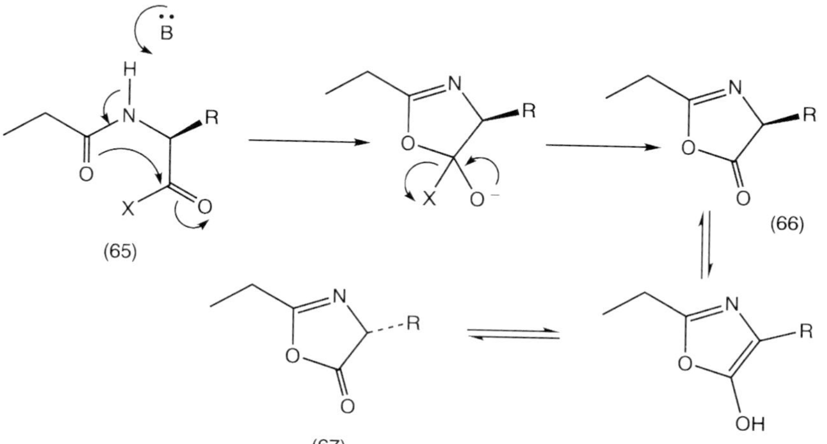

(65) (66) (67) (68)

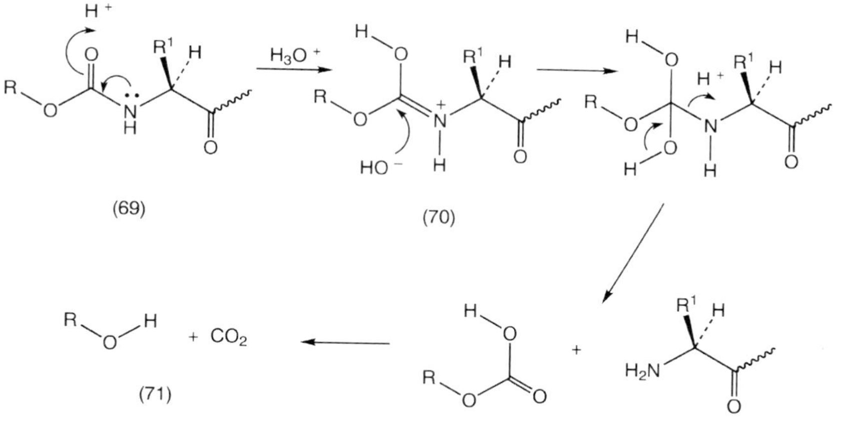

(69) (70) (71)

formed from the amine and the corresponding chloroformate in the presence of triethylamine. These groups may be cleaved with acid or with trimethylsilyl iodide.

The benzyloxycarbonyl (carbobenzyloxy) (Cbz) group was introduced by Bergmann in 1932.[202, 203] The reagent, benzylchloroformate (72), is obtained from benzyl alcohol and phosgene and it reacts with amines to give the benzyloxycarbonyl derivative (73). These protecting groups

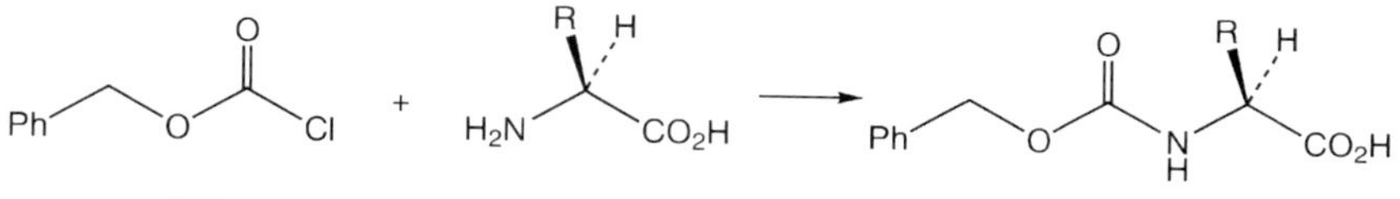

(72) (73)

can be removed by catalytic hydrogenolysis over palladium or by using sodium in liquid ammonia. This group is sensitive to treatment with hydrobromic acid in glacial acetic acid, conditions that make use of the ability of an aromatic ring to stabilise an adjacent radical or carbocation.

The sensitivity of the protecting group to these cleavage conditions may be altered by varying the substituents on the aromatic ring. For example a *para* nitro substituent facilitates hydrogenolysis presumably by stabilising a benzyl radical (**74**), whereas a *para* methoxyl group facilitates cleavage with trifluoroacetic acid at 0°C by stabilising a benzylic carbocation (**75**).[204, 205] In contrast the *p*-nitrobenzyloxycarbonyl group

(74)

(75)

is more stable to acid than the unsubstituted benzyloxycarbonyl group. This has been exploited in a peptide synthesis in which the more stable *p*-nitrobenzyloxycarbonyl group was used to protect the side-chain ε-amino group of lysine during preferential removal of the benzyloxycarbonyl group from the α-amino group.[206] Trimethylsilyl iodide can also be used to cleave this protecting group. The silicon readily bonds to the oxygen atoms, facilitating attack by the iodide ion. The N-hydroxybenzotriazole derivative (**76**) can be a useful mild way of introducing the benzyloxycarbonyl group.

(76)

The t-butyloxycarbonyl group (t-Boc) (**77**) is another frequently used protecting group.[207, 208] Deprotection under acidic conditions (HCl, CH_2Cl_2; CF_3CO_2H) gives the gases isobutene (**79**) and carbon dioxide. Anisole (**80**) can be added to the reaction mixture as a trap for the t-butyl carbocation (**78**) particularly when the substrate contains a reactive aromatic ring which might be alkylated. The protecting group was introduced as the t-butylazidoformate (**81**) in which the azide group is a good leaving group. However, this reagent is hazardous and alternative methods using di-t-butyldicarbonate (**82**) or the oxime derivative (**83**)

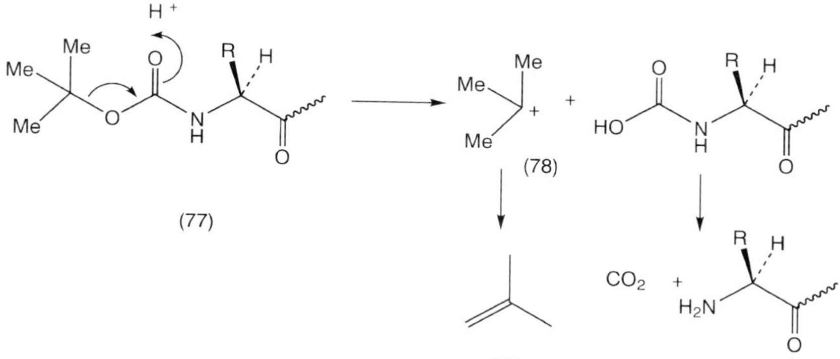

(77)

(78)

(79)

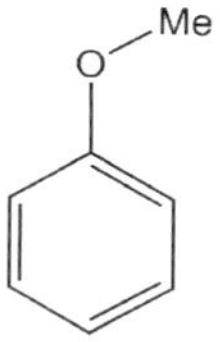

(80)

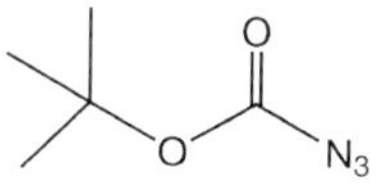

(81)

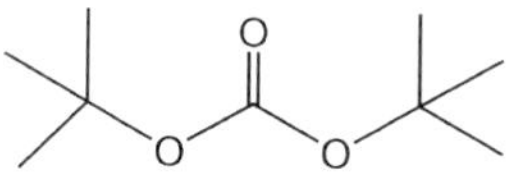

(82)

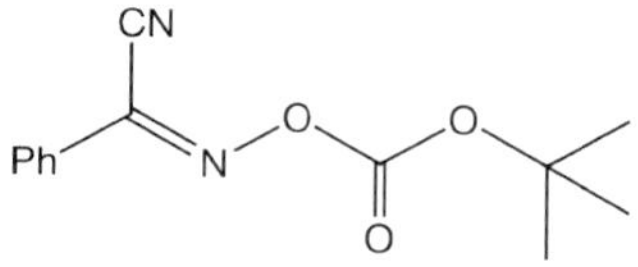

(83)

have been introduced.[209, 210] Whilst the former behaves as an anhydride, the reactivity of the latter is a result of the cyano grouping. A tetrahedral intermediate (85) is formed from (84). This intermediate breaks down via a ketenimine (86) releasing the protected amine (87).

The t-butyloxycarbonyl group is stable towards strong base and has been used in protecting arylamines during lithiation and functionalisation of the aromatic ring. The preparation of the protected derivative, which is

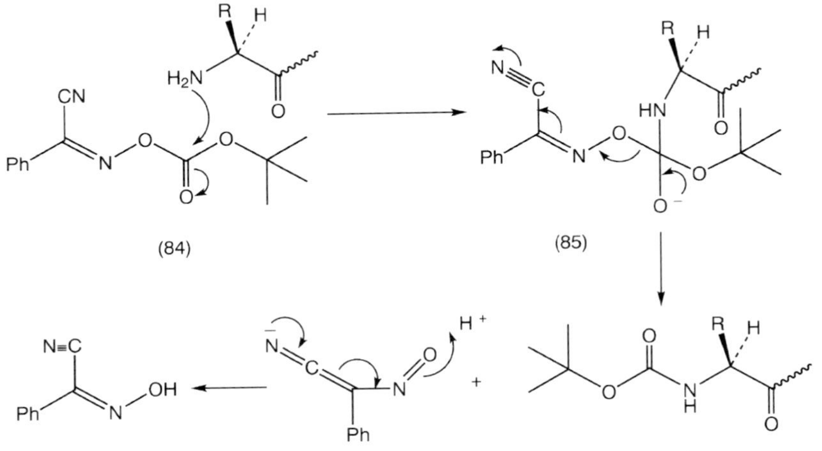

(84)   (85)

(86)   (87)

more difficult because of the decreased nuceophilicity of the arylamine, can be facilitated by first making the sodium derivative of the amine with sodium hexamethyldisilylazide.

A useful modification of the t-butyloxycarbonyl group is the diphenyl-oxycarbonyl (Bpoc) group (88) which utilises the ability of an aromatic ring to stabilise an adjacent carbocation (89) in the deprotection step. This protecting group can be removed using 2,2,2-trifluoro-ethanol as the proton source.

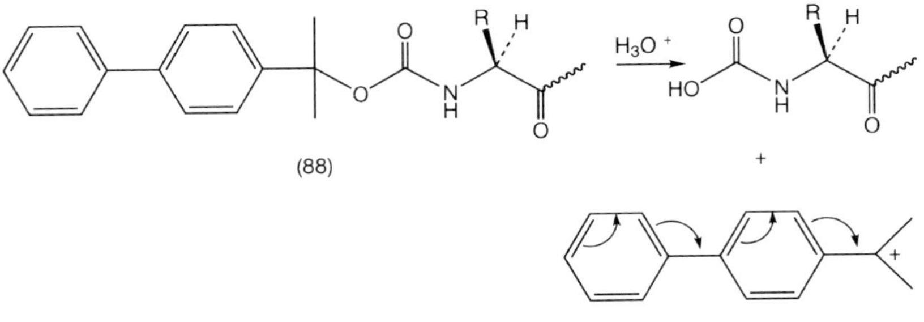

(88)

(89)

Whereas these protecting groups are removed under acidic conditions, the 9-fluorenylmethoxycarbonyl (Fmoc) group (90) is base labile.[212] The ester fragments in the presence of ammonia, piperidine or morpholine by an El_cB β-elimination which generates a cyclopentadiene anion that confers aromaticity on the central ring of the fluorene. The diphenyl-fulvene (91) which is formed, is trapped by the base, such as piperidine

(90)

(91)

or morpholine. This particular protecting group is acid stable, in contrast to a number of other amine protecting groups.

The 9-fluorenylmethoxycarbonyl group is usually introduced by re-acting the amine with 9-fluorenylmethyl chloroformate in the presence of base although a number of other Fmoc transfer reagents such as 9-fluorenylmethylpentafluorophenylcarbonate or N-(9-fluorenylmethoxy carbonyloxy)succinimide have been used. A variant of this protecting group is the tetrabenzo derivative (92) prepared via phenanthrene.[213]

(92)

The extra benzene rings give this compound an affinity for porous graphitised carbon through the interaction of the aromatic rings with the graphite. This allows the separation of peptides containing this protecting group attached, for example, to the terminal nitrogen of a peptide.

Fragmentation induced by carbanion formation is a further way of regenerating the amino group, which has been exploited with other protecting groups exemplified by (93)[214] and (94)[215] in which the sulfone and the nitrile, respectively, stabilise carbanion formation. The lability of these protecting groups meant that they were sometimes partially lost during the coupling process. Hence some two-step deprotection methods were devised in which the activation for the fragmentation was generated in the first step of the deprotection sequence. This is exemplified by the conversion of (95) to (93).

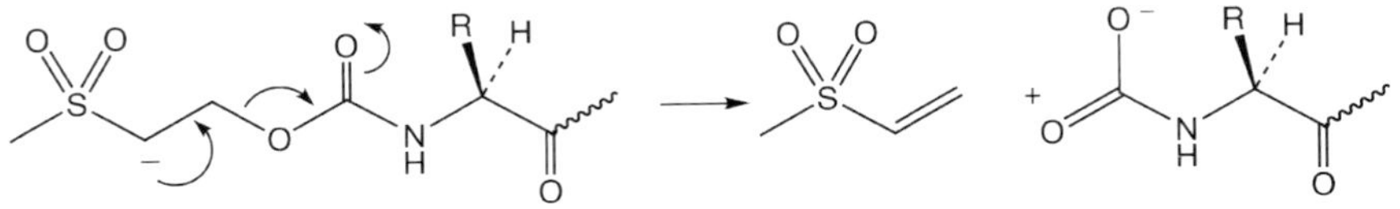

(93)

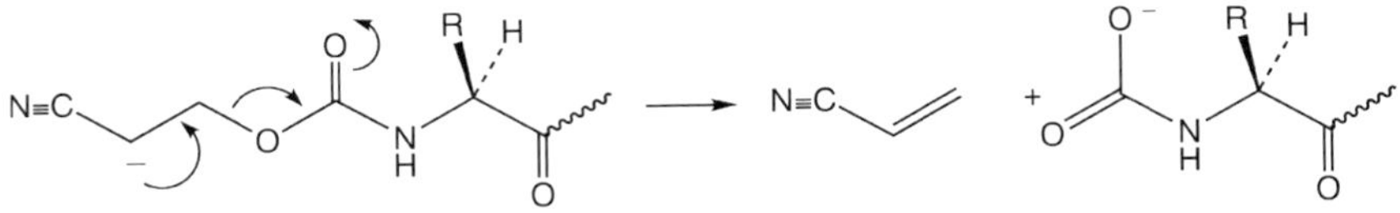

(94)

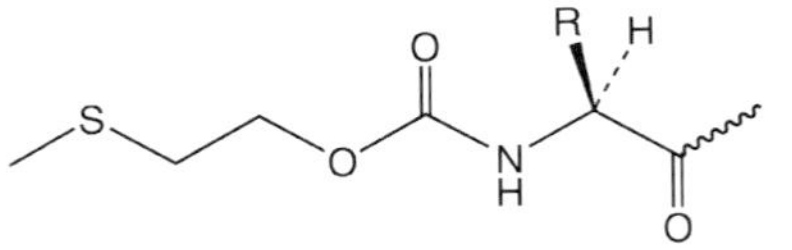

(95) 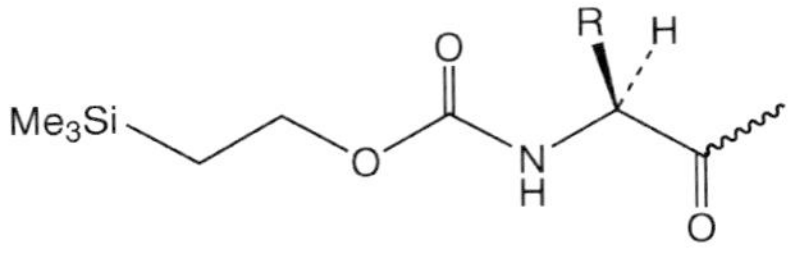   (96)

The 2-(trimethylsilyl)ethoxycarbonyl (96) protecting group is another group which can be removed by a carbanion-induced fragmentation.[216] The fluoride-ion cleavage of the trimethylsilyl group can generate a carbanion under relatively mild conditions. This protecting group is introduced using the corresponding chloroformate or N-succinimidyl carbonate.

Reductive processes using zinc may also initiate fragmentation of haloethoxycarbonyl protecting groups[217] such as the trichloroethoxycarbonyl protecting group (97).[47] The protecting group is introduced using trichloroethoxychloroformate and triethylamine.

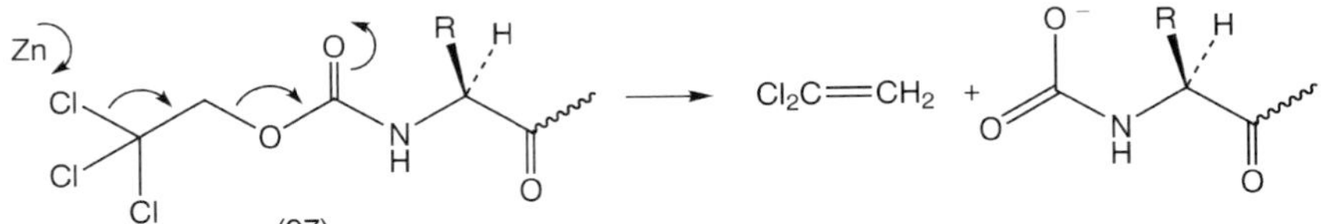

(97)

The opportunity for a catalyst to coordinate to neighbouring centres may also facilitate hydrolysis. For example a t-Boc protected amide (98) or carbamate undergoes deprotection under mild conditions with magnesium perchlorate in acetonitrile[218]—conditions where simple t-Boc protected amines are stable.

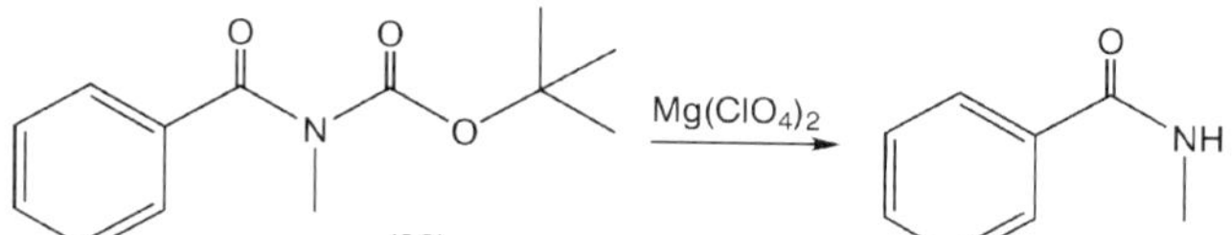

(98)

5.6 The protection of side chain amino groups in amino acids

The amino acid lysine contains a terminal amino group (**99**) and hence there is a problem of differentiating between two amino groups in the same molecule. This has been solved in a novel way.[219] The α-amino group participates in a copper chelate complex (**100**) which masks the amino group, leaving the terminal ε-amino group free for protection as, for example, its t-Boc derivative. The terminal guanidino group of arginine is protected as its nitro derivative (**101**)[220, 221] which substantially reduces its basicity. The nitro group may be removed by catalytic reduction or by an electrolytic reduction if a catalytic method is precluded by the presence, for example, of sulfur.

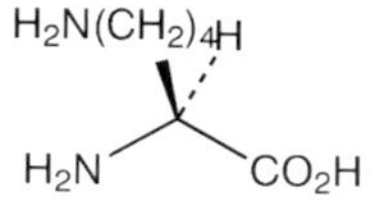

(99)

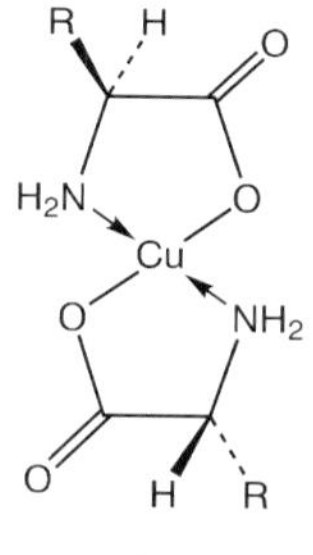

(100)

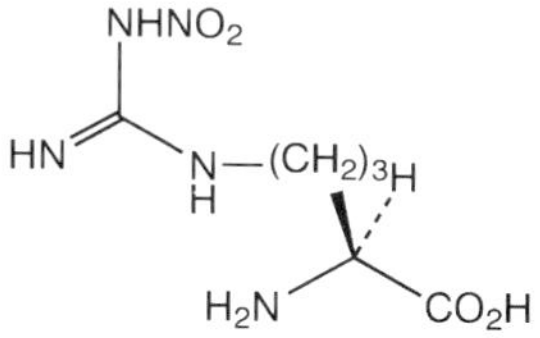

(101)

6 The protection of thiols

6.1 Introduction

The protection of a thiol is important because of the reactivity of the thiol as a nucleophile and its sensitivity to oxidation both by dimerisation and S-oxide formation. Although much of the effort in this area has been directed at the protection of the side chain of the amino acid cysteine (**1**),

(1)

most of the methods are general. Many of the principles associated with alcohol and amine protection apply to thiol protection. However, in the cleavage of thiol protecting groups use is made of the sensitivity of thioethers to cleavage by hydrogenolysis and by reaction with thiophilic heavy metals such as mercury, cadmium and nickel. The regiochemistry of this cleavage may be determined by the substituents in the protecting group.

6.2 Thioethers

The commonest method of thiol protection involves the formation of a thioether. The regiochemistry of the cleavage of the thioether is determined by its structure. Cleavage of simple thioalkyl ethers are not particularly regiospecific. However, the benzyl group, by stabilising carbocation formation, favours deprotection to leave the sulfur in place in the substrate. S-Benzyl ethers (**2**)[222] are formed by reaction with benzyl

(2)

chloride in the presence of a base and are cleaved by hydrogenolysis with sodium in liquid ammonia[222-224] or by treatment with hydrogen fluoride.[225] Since benzyl fluoride is formed in the latter, the aromatic ether, anisole, is often used as a cation scavenger to prevent unwanted acid-catalysed side reactions.

Modified benzyl thioethers such as *p*-methoxybenzylthioether (3)[226] have been introduced since they are more readily cleaved by mixtures of

(3)

acids such as anhydrous trifluoroacetic acid containing phenol or anisole. The di- and triphenylmethyl thioethers (4)[227] are useful not only because

(4)

they are more readily cleaved by acid but also because steric factors reduce the propensity of the sulfur to be oxidised to sulfoxides. The triphenylmethyl thioethers (4) may be prepared by using triphenylmethyl chloride. The difference between the sensitivity of ethers and esters to acid and base cleavage which has been utilised in the protection of alcohols has parallels in thiol protection. Mercury(II) acetate has been found[227] to cleave selectively the S-triphenylmethyl group from a cysteinyl peptide, whereas the base sodium ethoxide was used to remove selectively the S-benzoyl group.

6.3   The protection of thiols as acetal derivatives

Some of the principles that facilitate the cleavage of acetals and which were utilised in the development of methoxymethyl ethers have been introduced into the protection of thiols. The S-methoxymethylmonothio-acetal (5) protecting group may be introduced by using dimethoxy-methane[228] or with the toxic chloromethylmethyl ether. The cleavage

(5)

of these can be achieved by using Lewis acids or by simple acid catalysis. Dithioacetals are cleaved by heavy metals and use can be made of this with the S-benzylthiomethyl (6) protecting group. This is intro-

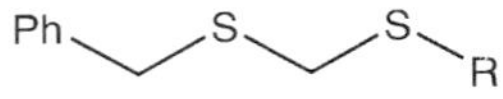

(6)

duced[229, 230] by using chloromethylbenzylsulfide in the presence of base and it is cleaved by reaction with mercury(II) acetate.

The participation of an acetamido grouping in the acid-catalysed cleavage of a thioacetal has been used[231] in the S-acetamidomethyl (7)

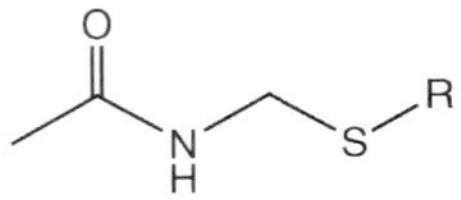

(7)

protecting group. Protonation of the amide function facilitates the cleavage of this grouping.

The 9-fluorenylmethyl (8) group, which has found application in the protection of amines, is also used to protect thiols.[232] It is acid stable, and

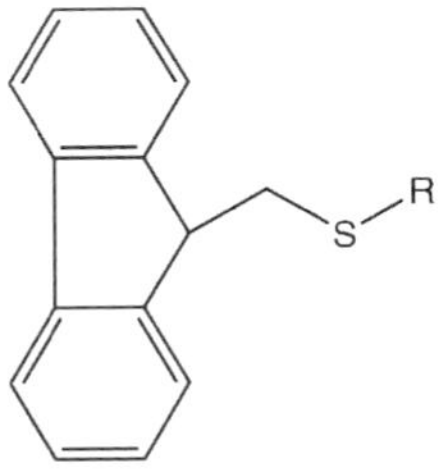

(8)

ammonia or piperidine in dimethylformamide are used as the base in deprotection. The use of carbanion-induced fragmentation is also used in the removal of other thiol protecting groups, for example the cyanoethylthio ether (9).[233] The introduction of this protecting group is

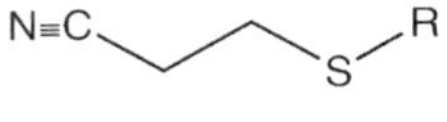

(9)

facilitated by the ability of a thiol to act as a nucleophile in a Michael addition reaction to acrylonitrile. Deprotection was carried out with potassium t-butoxide.

The silicon–sulfur bond is considerably weaker than the silicon–oxygen bond and therefore silyl thioethers are more easily cleaved than silyl ethers. The application of silylthioethers as protecting groups for thiols is therefore limited.

6.4 Thioesters as protecting groups

Although perhaps thioesters are better known because of the reactivity confered by the sulfur, they are in principle available as protecting groups. Thiocarbonates have been described as protecting groups for thiols. The t-butyloxycarbonyl (**10**) and benzyloxycarbonyl (**11**) groups

(10)   (11)

are introduced via the corresponding chloroformates.[234, 235] They have the useful property of reducing the nucleophilicity of the sulfur. There is some limited selectivity in their cleavage by acid compared with the corresponding carbamates which are hydrolysed more readily.

Thiols are sometimes protected as their symmetrical dithiols. Thus the dimer, cystine, may be used as a protected form of cysteine. An unsymmetrical disulfide protecting group is afforded by the o-nitrophenyldithio ether (**12**). The latter is introduced by using o-nitrobenzene-

(12)

sulfenyl chloride, a reagent more commonly found in the protection of amines. This protecting group is cleaved reductively, for example with sodium borohydride or by treatment with excess mercaptoethanol. o-Nitrophenylsulfenyl chloride may also be used to remove a triphenylmethyl or tertiarybutyl protecting group from the sulfur of cysteine through the formation and reductive cleavage of a dithioether.[236, 237]

7 The protection of unsaturated systems

7.1 The protection of alkenes

Although selectivity between double bonds and other reactive centres in a molecule can generally be achieved by the careful choice of reagents, there are a number of situations in which the protection of the double bond is required. This can be the case when vigorous reaction conditions are involved and double-bond isomerisation, skeletal rearrangement or allylic oxidation are possible.

An alkene may be protected as its dibromide. The stereospecificity of the addition of bromine and the method to be used for the subsequent regeneration of the alkene are important considerations if the stereochemistry of the original double bond is to be conserved through the sequence. Dibromides may be formed by using bromine in glacial acetic acid or, more conveniently on the small scale, pyridinium hydrobromide perbromide.[238] The double bond is regenerated by using metals such as zinc or low-valency metal ions such as chromium(II) or tin(II). A classic example[239] of the use of a dibromide in protecting an alkene is shown by the oxidation of cholesteryl acetate (1) with chromium trioxide. A range of products are formed by attack on the double bond and at the allylic position, exemplified by (2). However, if the double bond is protected as the dibromide (3), the side chain is cleaved and (4) is obtained. The double bond (5) is regenerated by elimination with zinc as in (4) or with sodium iodide. The procedure may be further adapted to afford the 14α-hydroxy derivative (6).[240] Dichlorides[241] or halohydrins may be used for a similar purpose but care must be taken to see that the original geometry of the alkene is restored.[242]

An epoxide can be used to protect a double bond. The plant hormone, gibberellic acid (7), possesses two double bonds and it is possible to distinguish between them. Epoxidation of the 13-monoacetate of the methyl ester with m-chloroperbenzoic acid gave (8), which was used[243] to protect the 16,17-double bond selectively and allow reduction of the ring-A double bond to give (9). Removal of the epoxide from (9) to afford (10) was achieved by reduction with sodium iodide, sodium acetate, glacial acetic acid and zinc dust.[244] The other protecting groups were then removed to give gibberellin A_1 (11). Double bonds may also be regenerated from epoxides by reduction with triphenylphosphine, trimethylphosphite, chromium(II) chloride, a low-valency state of

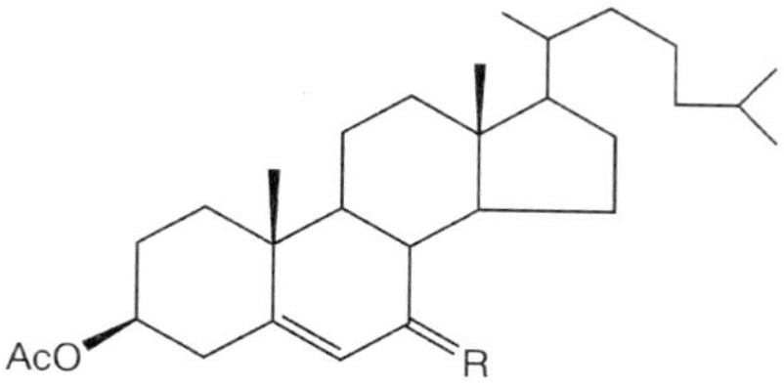

(1) R = H$_2$
(2) R = O

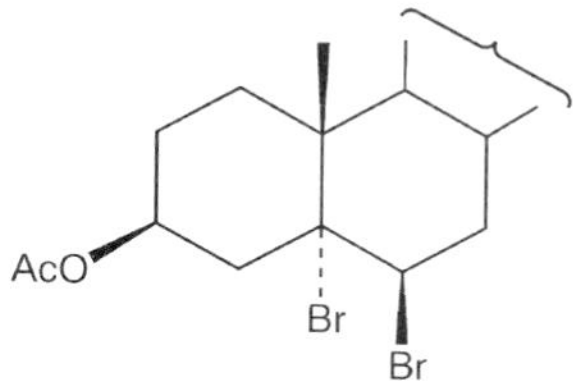

(3)

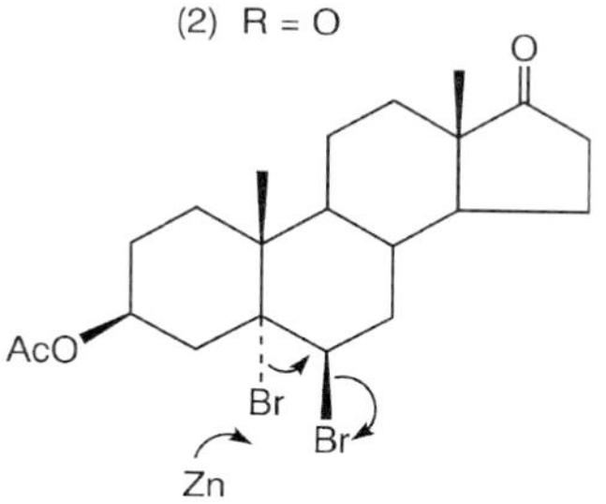

(4)

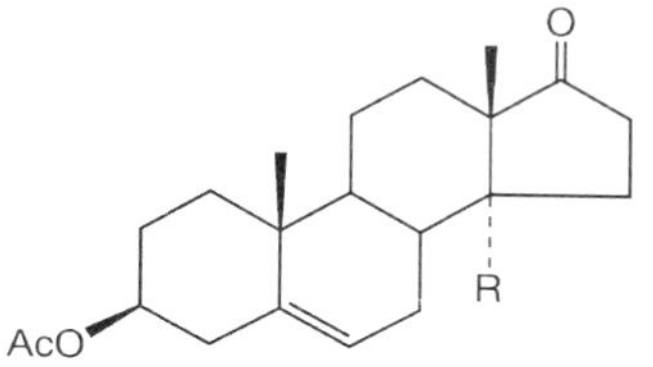

(5) R = H
(6) R = OH

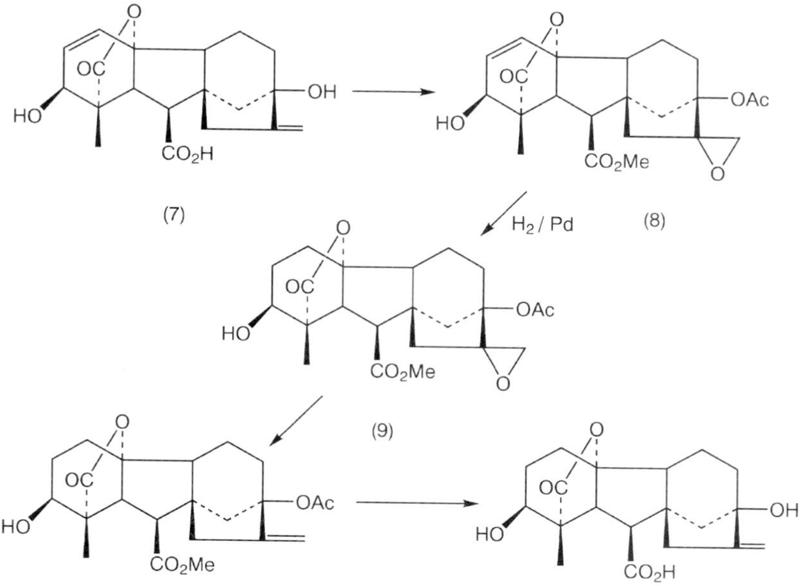

(7)

(8)

(9)

H$_2$/ Pd

(10)

(11)

tungsten or with toluene-*p*-sulfonic acid and sodium iodide in dry acetonitrile.[245]

A further way of protecting an alkene is to convert it to a *cis* or *trans* 1,2-diol (**12**) by using either osmylation or acid-catalysed hydration of an epoxide. The alkene (**14**) may be regenerated from the *cis* diol via the 2-thio-1,3-dioxolane (**13**) and reduction with Raney nickel or a trialkyl-phosphite.[246, 247] The 2-thio-1,3-dioxolanes can be prepared from vicinal

| (12) | (13) | (14) |

diols by using thiophosgene, N,N′-thiocarbonyldiimidazole or 1,1-thiocarbonyldi-2,2′-pyridone.[248] Dimethylamino-1,3-dioxolanes can be prepared from *cis* diols by using dimethylformamide dimethylacetal and converted to alkenes with acetic anhydride.[249] In the case of *trans* diols, reduction of the dimethanesulfonate with zinc may be used to regenerate the alkene.

7.2   The protection of dienes

Diels–Alder reactions can be reversible and hence Diels–Alder adducts may be used to protect some alkenes which act as dienophiles and also to protect selected dienes. Thus the mono-adduct (**17**) of the quinone (**15**) and the cyclopentadiene (**16**) was modified, and then the alkene (**18**) was reformed by a reverse Diels–Alder reaction.[250]

| (15) | (16) | (17) | (18) |

The adduct (21) between the diene of ergosteryl acetate (19) and 4-phenyl-1,2,4-triazoline-3,5-dione (20) has been used to protect the ring-B diene while the side chain of ergosterol was cleaved by ozonolysis.[251, 252] An efficient synthesis of 1α,25-dihydroxyvitamin D_3 (23), which

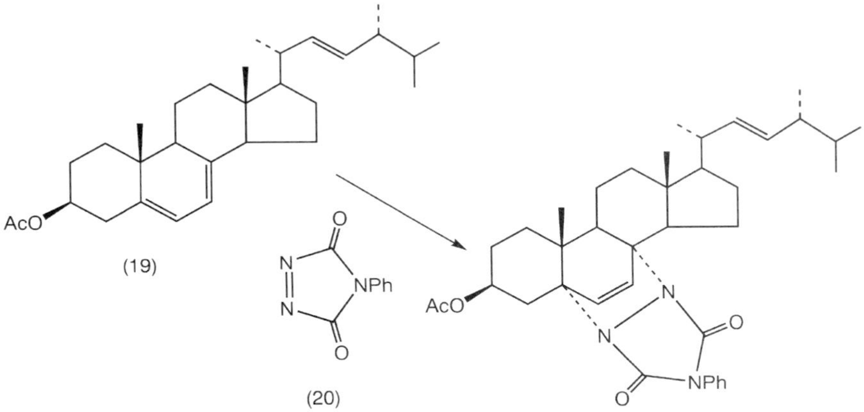

(19)

(20)

(21)

is the biologically active metabolite of vitamin D_3, utilised the 1,2,4-triazoline adduct (22) in which the triene was regioselectively blocked, allowing allylic oxidation of the ring-A double bond.[253] Metal complexes

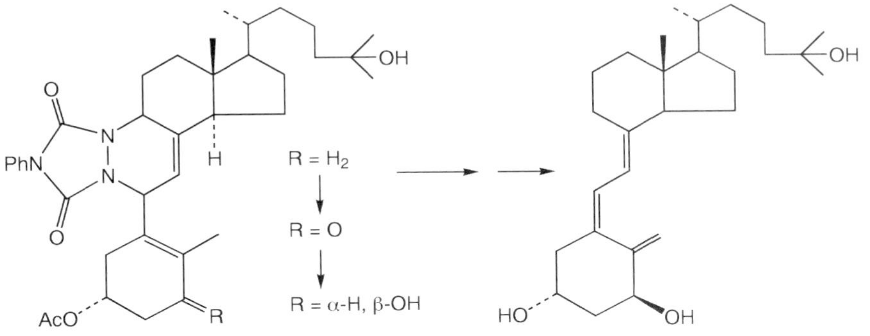

(22)

(23)

(for example 24) using iron pentacarbonyl can also be prepared from the ring-B diene of ergosterol, allowing the side-chain double bond to be modified by reduction, hydroboration or osmylation. The diene may be obtained from the complex by oxidation with iron(III) or ceric salts.[252]

(24)

The extrusion of sulfur dioxide from a cyclic sulfone may give a diene and, since sulfur dioxide may also add to dienes, this fulfils some of the criteria of protection. An example of sulfur dioxide being used in this way comes from a short synthesis of the yohimbine alkaloids (25–27).[254] The diene in (26), formed by the extrusion of sulfur dioxide from (25), undergoes an internal Diels–Alder cyclisation to afford (27), which was then transformed into (28).

7.3   The protection of alkynes

The protection of alkynes presents a difficult problem because, although the triple bond readily adds reagents such as bromine, the regiospecific regeneration of the triple bond free from isomeric impurities, such as allenes, is difficult. A solution to this problem involves complexation with dicobalt octacarbonyl (for example in **29–30**).[255, 256] An example in a synthesis is given in the sequence (**31**)–(**34**). The acetylene in (**34**) was regenerated by oxidation of the metal with iron(III) nitrate. Ceric ammonium nitrate or trimethylamine N-oxide are alternative oxidants.

The acetylenic hydrogen is sufficiently acidic to require protection, for example as its trimethylsilyl derivative (**35**).[257, 258]

7.4   Positional protective groups in aromatic substitution

Electrophilic aromatic substitution is governed by well-established rules. The orientation of the product of electrophilic attack is determined by the incoming electrophile reacting at the most accessible site of highest electron density. This outcome may be modified by first blocking this site with a group that can be subsequently removed. A number of groups have been used for this purpose. Typical groups are the sulfonoxy, nitro, carboxyl, trichloroacetyl and halogen substituents. An amino group can also be used as a directing group and then removed by diazotisation and reduction of the diazo compound with hypophosphorus acid. The following examples illustrate the use of these groups.

Sulfonation of acetanilide (36) gave the *para*-sulfonic acid (37) which was nitrated *ortho* to the acetanilide (38). Removal of the sulfonic acid

from (38) by acid hydrolysis afforded 2-nitroaniline (39).[259] 2,6-Dinitroaniline (44) can be prepared from chlorobenzene (40) by sulfonation (41) and dinitration to afford (42). The chlorine in (42) is then susceptible to nucleophilic displacement with ammonia to give (43), and the sulfonic acid grouping is finally removed by acid hydrolysis to afford (44).[260]

An example of the use of a carboxyl group in protecting an aromatic site in synthesis is shown by the preparation of some 2-substituted 3,4-xylenols.[261] Carboxylation of 3,4-xylenol (45) afforded (46), which was then brominated in the 2-position to give (47). Decarboxylation of the

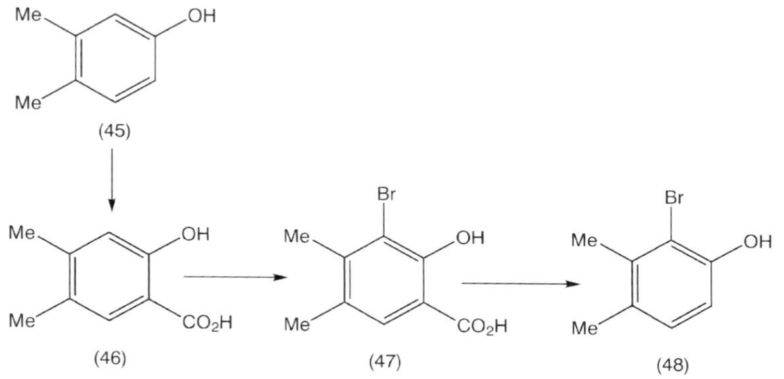

(45)

(46)   (47)   (48)

phenolic acid by heating (47) in quinoline gave the required product (48). In considering this decarboxylation it should be noted that the phenol (47) represents the enolic form of a β-keto-acid.

The use of an amino group to protect a position is revealed in the synthesis of a *meta*-substituted alkylbenzene, for example that of *m*-nitrotoluene (53) which is shown in the sequence (49)–(53).[262]

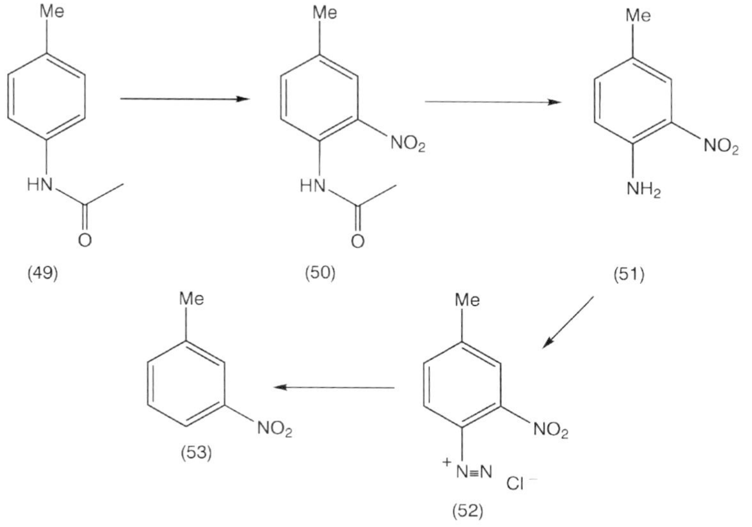

(49)   (50)   (51)

(53)

(52)

The synthesis of the alkaloid (56) illustrates the use of bromine to protect a position *para* to the phenol in (55). The Mannich-type reaction

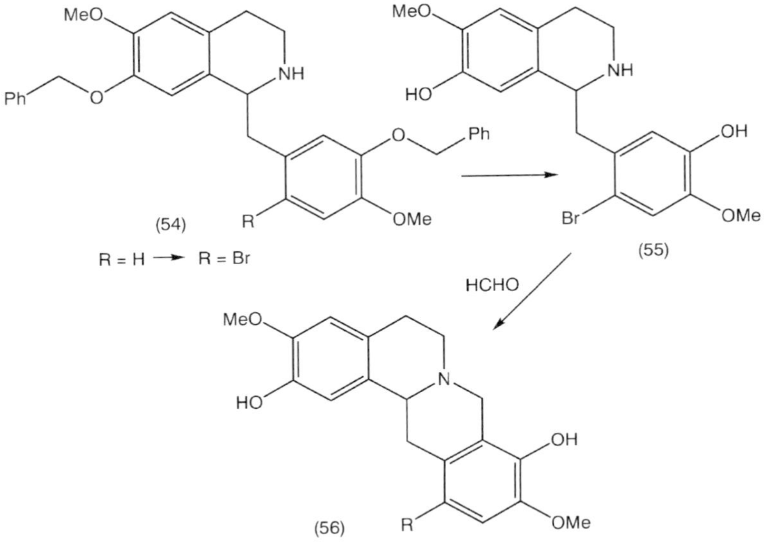

(54)

R = H ⟶ R = Br

(55)

(56) R = Br ⟶ R = H

which was used to construct the 'berberine' bridge was directed to follow the route from (54) to (56). Reductive removal of the halogen gave the product.[263]

Reductive dehalogenation of bromo- and iodophenols proceeds selectively, and it is possible to use the bromo and iodo substituents as

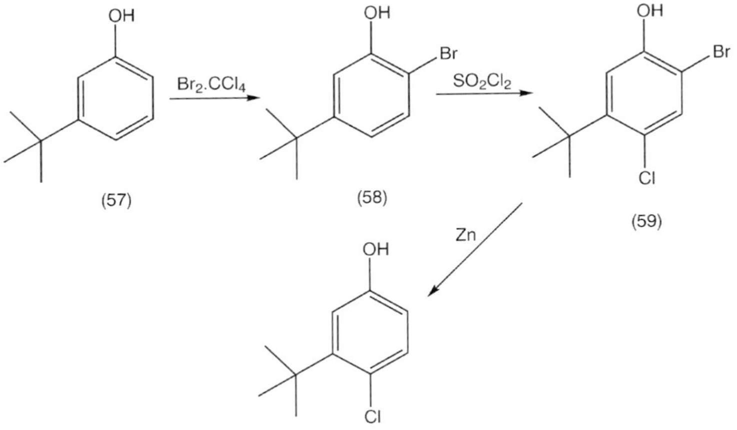

(57) (58) (59) (60)

positional protective groups. 4-Chloro-t-butylphenol (60) is a powerful fungistatic agent.[264] Direct halogenation of *m*-t-butylphenol (57) gives the *ortho*-halogenated product (58) and little of the *para* product (60). Bromination and iodination of (57) gave the 2-bromo- and 2-iodo-5-t-butyl-phenols in high yield. Chlorination with sulfuryl chloride then afforded the 4-chloro compound (59) from which the bromine or iodine protecting atoms were removed by reduction with zinc in acetic acid.

8 Some experimental methods

This chapter is concerned with some common general experimental methods for introducing protecting groups. One of the major criteria for a good protecting group is that it should be introduced under as mild conditions as possible. Since the steric and electronic environment of a functional group may vary widely from one compound to another, general procedures need to be modified to suit the particular substrate. The selectivity that is often required for the introduction of a particular protecting group may often only be achieved after careful preliminary experimentation. The general procedure may need to be adapted to the specific case by following the reaction by thin-layer chromatography and by varying the reaction time and temperature.

It is important to emphasise that many of the reagents are highly reactive and in some instances, such as diazomethane, are powerful alkylating agents and are consequently toxic. Attention must be paid to the safety aspects of these reactions by using well-ventilated fume cupboards and by carefully considering the method of disposal of excess reagent. It is important to draw up the individual hazard and risk assessment for the protection and deprotection stages of a reaction sequence as separate steps.

The high reactivity of many of the reagents also means that they may need to be purified immediately prior to use. They may also be sensitive to impurities in the solvent. Pyridine, for example, is often used as a solvent. Many acylations, which can be carried out in pyridine, are particularly sensitive to traces of water and thus it is important to purify the pyridine by, for example, distillation from solid potassium hydroxide and storing it over a Linde type 4A molecular sieve or calcium hydride. Tetrahydrofuran is another example of a common solvent which needs to be dried carefully by, for example, distillation from lithium aluminium hydride and storage over calcium hydride. For further details the reader is referred to standard books on the purification of organic reagents such as *Purification of Laboratory Chemicals* by D.D. Perrin, W.L.F. Armarego and D.R. Perrin.[265]

The following notes are of general guidance for the introduction of some of the more common protecting groups. Many of the procedures have been used in the author's laboratory. Further collections of experimental procedures may be found in the recent editions of Vogel's *Textbook of Practical Organic Chemistry*[266] and in the various preparations given in *Organic Syntheses*.[267]

8.1 The protection of alcohols

The various protecting groups for alcohols have been described in chapter 2. Many of the reagents that are used in the protection of alcohols are sensitive to the presence of traces of water and hence it is important to exclude water not only by drying the solvent but also by ensuring that the alcohol itself is dry. Some of the reactions generate hydrochloric acid which can react with the substrate. Hence a base may be used not only to catalyse the reaction but also to react with hydrochloric acid. Sometimes the last traces of pyridine may be difficult to remove from the reaction product even though the work-up may have involved treatment with dilute hydrochloric acid. A rinse with aqueous copper sulfate, which forms a complex with pyridine, may be helpful.

8.1.1 The preparation of a methyl ether

8.1.1.1 Preparation of silver oxide
Silver nitrate (17 g) was dissolved in distilled water (170 cm^3) and the solution was warmed to 85°C in a flask protected from the light by black paper. A solution of sodium hydroxide (3.9 g) in water (170 cm^3) was added rapidly and the mixture was stirred vigorously for 15 minutes. The mixture was allowed to cool and stand until the coagulation of the precipitated silver oxide was complete. The silver oxide was collected with use of a Buchner flask and funnel with protection from direct light. The precipitate was washed thoroughly with water and a little acetone and ether before being dried under high vacuum.

8.1.1.2 Methylation
The substrate (17β-acetoxy-3β-hydroxyandrost-4-ene, **1**) (1 g) in dry dimethylformamide (20 cm^3) was treated overnight with silver oxide (1.6 g) and methyl iodide (4 cm^3) with thorough stirring and protection from the light. The reaction mixture was filtered through a Celite pad. The pad was washed with ether. The solvents were evaporated *in vacuo* and the product was chromatographed on silica. Elution with 2% ethyl acetate in light petroleum gave the product (17β-acetoxy-3β-methoxy-androst-4-ene, **2**) (700 mg) which crystallised from light petroleum as needles, melting point (m.p.) 107°C–110°C.

8.1.2 The preparation of a tetrahydropyranyl ether

The substrate (17β-acetoxyandrostan-4β,5α-diol, **3**) (1 g) in dry diethyl ether (25 cm^3) was treated with dihydropyran (0.5 g) and toluene-*p*-sulfonic acid (50 mg) at room temperature for 12 h. The mixture was

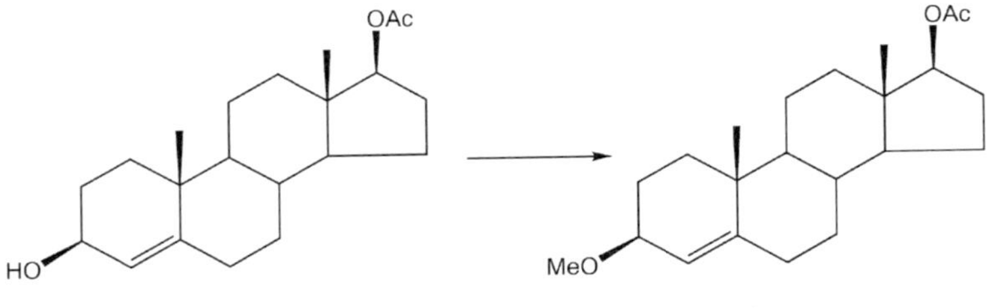

(1)   (2)

diluted with diethyl ether, washed with aqueous sodium hydrogen carbonate, brine and dried over sodium sulfate. The solvent was evaporated to give a residue which was chromatographed on silica. Elution with 10% ethyl acetate:light petroleum gave the product (17β-acetoxy-4β-tetrahydropyranyloxy-androstan-5α-ol, **4**) (0.9 g) which crystallised from ethyl acetate:light petroleum as prisms, m.p. 149°C–153°C.

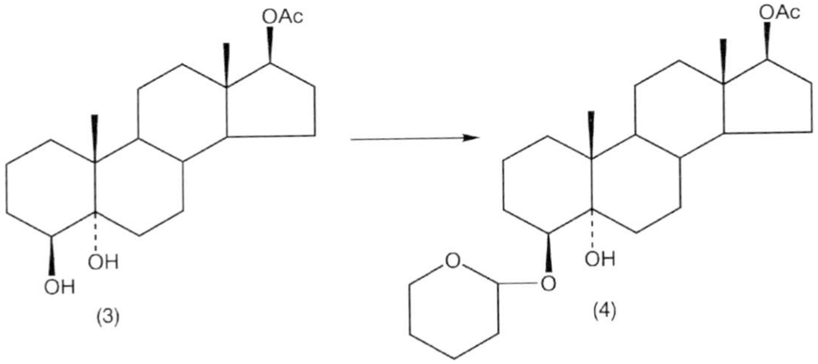

(3)   (4)

8.1.3 The preparation of a methoxymethyl ether

The substrate (3β,4β-dihydroxyandrost-5-en-17-one, **5**) (1 g) in dry dichloromethane (25 cm^3) was treated with N,N-diisopropylethylamine (3 g) and methoxyethoxymethyl chloride (3 g) at room temperature for

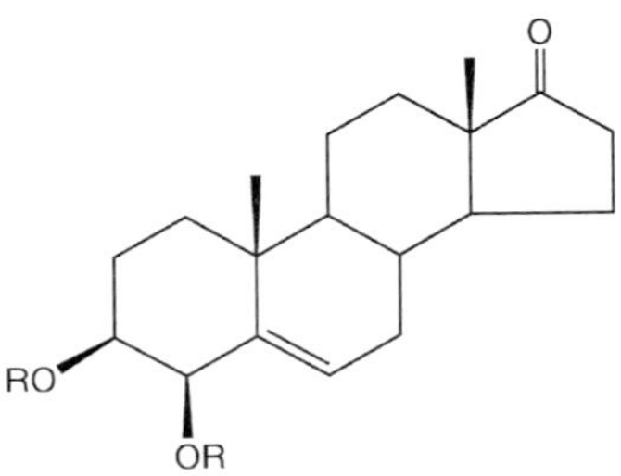

(5) R = H

(6) R =

24 h. The mixture was poured into dilute hydrochloric acid and the steroid was recovered in dichloromethane. The extract was washed thoroughly with water, aqueous sodium hydrogen carbonate, brine and dried over sodium sulfate. The solvent was evaporated and the residue was chromatographed on silica. Elution with 20% ethyl acetate:light petroleum gave the bismethoxyethoxymethyl ether (6) (0.7 g) as an oil lacking hydroxyl absorption in the infrared spectrum.

8.1.4   The preparation of a triphenylmethyl ether

The substrate (ent-6β,7α,18-trihydroxykaur-16-en-19-oic acid 19-6β-lactone, 7) (6 g) in dry pyridine (50 cm^3) was heated with triphenylmethyl

(7)   R = H

(8)   R = Ph$_3$C

chloride (6 g) for 8 h on a boiling water bath. The mixture was cooled and poured into dilute hydrochloric acid. The product was recovered in ethyl acetate. The extract was washed with water and dried over sodium sulfate. The solvent was evaporated to give a residue which was chromatographed on silica. Elution with 12.5% ethyl acetate:light petroleum gave the 18-triphenylmethyl ether of ent-6β,7α,18-trihydroxy-kaur-16-en-19-oic acid 19-6β-lactone (8) (5 g) which crystallised from ethyl acetate:light petroleum as needles, m.p. 209°C–211°C.

8.1.5   Methylation of a phenol by using dimethyl sulfate

The substrate (podocarpic acid, 9) (4 g) was suspended in a solution of sodium hydroxide (5 g) in water (20 cm^3) which was then cooled in an ice

(9)   R = H

(10) R = Me

bath. Dimethyl sulfate ($12\,cm^3$) was added dropwise to the solution which was then heated under reflux for 2 h. The solution was cooled, diluted with water and extracted with diethyl ether. The extract was washed with dilute hydrochloric acid, water and brine and dried over sodium sulfate. The solvent was evaporated and the residue was chromatographed on silica. Elution with light petroleum gave the product (methyl O-methyl-podocarpate, **10**) (2.5 g) which crystallised from aqueous ethanol as needles, m.p. 129°C–130°C.

8.1.6 Methylation of a phenol by using methyl toluene-p-sulfonate

The substrate (estrone, **11**) (1 g) and methyl toluene-p-sulfonate (1.5 g) were heated on a boiling water bath in 10% aqueous potassium hydroxide

(11) R = H
(12) R = Me

($10\,cm^3$) for 1 h under nitrogen. A further portion of 10% aqueous potassium hydroxide was added and the solution was heated for a further hour. The mixture was cooled and the methyl ether (**12**) (0.8 g) was collected and recrystallised from aqueous ethanol as needles, m.p. 168°C–169°C.

8.1.7 The preparation of a t-butyldimethylsilyl ether

t-Butyldimethylsilyl chloride (3 g) and imidazole (3.5 g) were added to a solution of the substrate (17β-acetoxy-3β-hydroxyandrost-4-ene, **1**) (2.5 g) in dry dimethylformamide ($25\,cm^3$). The mixture was stirred at room temperature overnight and then diluted with ether. The organic phase was washed with brine and dried over sodium sulfate. The organic solvents were evaporated to give a waxy solid which was recrystallised from acetone:light petroleum to give the product (17β-acetoxy-3β-t-butyldimethylsilyloxyandrost-4-ene, **13**) (3 g) as needles, m.p. 116°C–118°C.

8.1.8 Hydrolysis of a t-butyldimethylsilyl ether

A solution of the silyl ether (17β-acetoxy-4α,5α-epoxy-3β-t-butyldimethylsilyloxyandrostane, **14**) (2.4 g) in dry tetrahydrofuran ($20\,cm^3$)

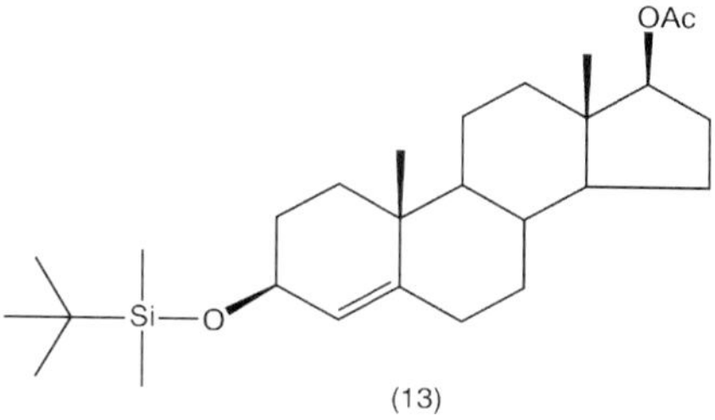

(13)

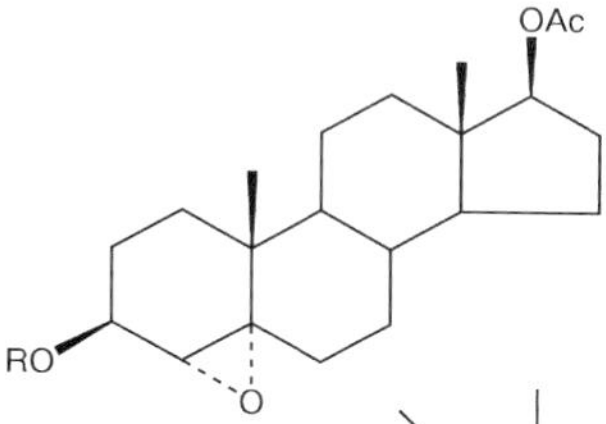

(14) R = [silyl group]

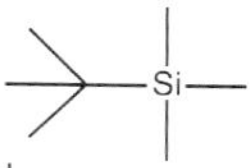

(15) R = H

was treated with a 1M solution of tetrabutylammonium fluoride in tetrahydrofuran (10 cm^3) for 3 h at room temperature. The solution was diluted with ethyl acetate (150 cm^3) and thoroughly washed with water. The organic phase was dried over sodium sulfate and the solvent evaporated to give a residue which was chromatographed on silica. Elution with 20% ethyl acetate:light petroleum gave the product (17β-acetoxy-4α,5α-epoxyandrostan-3β-ol, **15**) (1.25 g) which crystallised from ethyl acetate:light petroleum as plates, m.p. 126°C–127°C.

8.1.9 The preparation of an acetate ester

The substrate (17β-hydroxyandrost-4-en-3-one, **16**) (2 g) in dry pyridine (10 cm^3) was treated with acetic anhydride (4 cm^3) in a stoppered conical

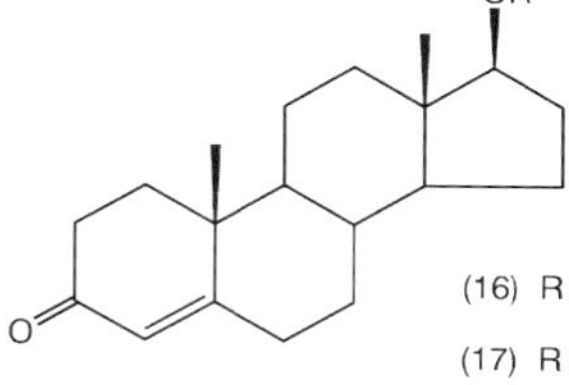

(16) R = H

(17) R = Ac

flask at room temperature overnight. The mixture was poured into excess dilute hydrochloric acid and the steroid extracted with ethyl acetate. The extract was washed with water, aqueous sodium hydrogen carbonate, brine and dried over sodium sulfate. The solvent was evaporated to give the product (17β-acetoxyandrost-4-en-3-one, **17**) (2 g) which crystallised from light petroleum as needles, m.p. 143°C.

8.1.10   Hydrolysis of an acetate ester

The substrate (3β,4β-diacetoxy-5α-androstan-17-one, **18**) (2 g) in methanol (50 cm^3) was treated with saturated aqueous sodium hydroxide

(18) R = Ac
(19) R = H

(2 cm^3) at room temperature for 1 h. Glacial acetic acid (4 cm^3) was added and the solution was concentrated *in vacuo* and poured into water. The steroid was recovered in ethyl acetate and the extract was washed with aqueous sodium hydrogen carbonate and dried over sodium sulfate. The solvent was evaporated to give the product (3β,4β-dihydroxy-5α-androstan-17-one, **19**) (1.3 g) which crystallised from methanol as plates, m.p. 221°C–222°C.

8.1.11   The preparation of an isopropylidene derivative

The diol (5β,6β-epoxy-3β,4β-dihydroxyandrostan-17-one, **20**) (1 g) in acetone (25 cm^3) was treated with concentrated sulfuric acid (0.1 cm^3) at

(20)

room temperature for 12 h. Excess aqueous sodium hydrogen carbonate was added and the product was collected by filtration to afford the derivative (5β,6β-epoxy-3β,4β-isopropylidenedioxy-androstan-17-one, **21**)

(869 mg) which crystallised from acetone as needles, m.p. 229°C–230°C.

(21)

8.2   The protection of aldehydes and ketones

The chemistry underlying the protection of the carbonyl group has been discussed in chapter 3. Many of the reactions involved in the protection of carbonyl groups are acid-catalysed and consequently it is important to remove the acid catalyst as early and as completely as possible in the work-up procedure to avoid loss of the protecting group. Where an aqueous work-up has to be avoided, the use of solid-phase Lewis-acid catalysts, acidic Montmorollinite clays or ion-exchange resins as the catalysts or alumina to remove the acid may be effective. It is also worth remembering that sodium hydrogen sulfate can be used as an acidic catalyst. A number of reactions also involve the loss of a molecule of water. This may be facilitated by carrying out the reaction in the presence of a mild dehydrating agent such as anhydrous copper sulfate or by azeotropically removing the water with use of toluene (less hazardous than benzene) in a Dean and Starke apparatus.

8.2.1   The preparation of a 1,3-dioxolane derivative

A solution of the substrate (ent-16-oxobeyeran-19-oic acid, **22**) (917 mg) and toluene-*p*-sulfonic acid (40 mg) in toluene (100 cm^3) and ethylene glycol (40 cm^3) was subjected to azeotropic distillation, by using a Dean and Starke apparatus, for 5 h. Fresh toluene was added at regular intervals to maintain the original volume. The mixture was cooled and diluted with ethyl acetate. The solution was washed with water, brine and dried over sodium sulfate. The solvent was evaporated *in vacuo* to afford the derivative (the 16-ethylene ketal of 16-oxo-ent-beyeran-19-oic acid, **23**) (890 mg) which crystallised from ethyl acetate as cubes, m.p. 255°C–258°C.

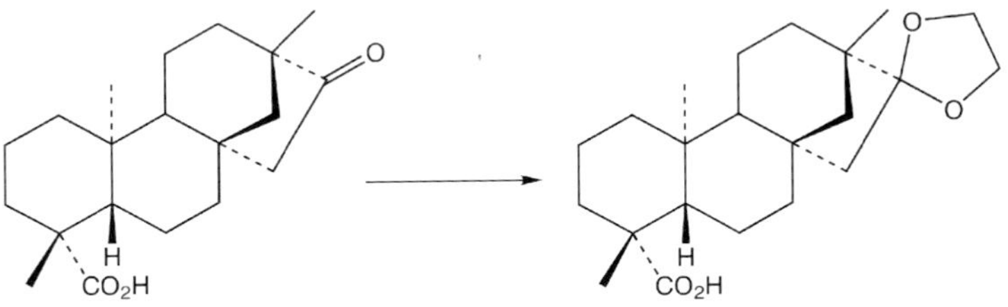

(22)   (23)

Deprotection was achieved by heating the acetal under reflux in ethanol containing a drop of concentrated hydrochloric acid.

8.2.2   The preparation of an oxime

A solution of the substrate (3β-hydroxy-5α-androstan-17-one, **24**) (3 g) and hydroxylamine hydrochloride (960 mg) in ethanol (30 cm^3) contain-

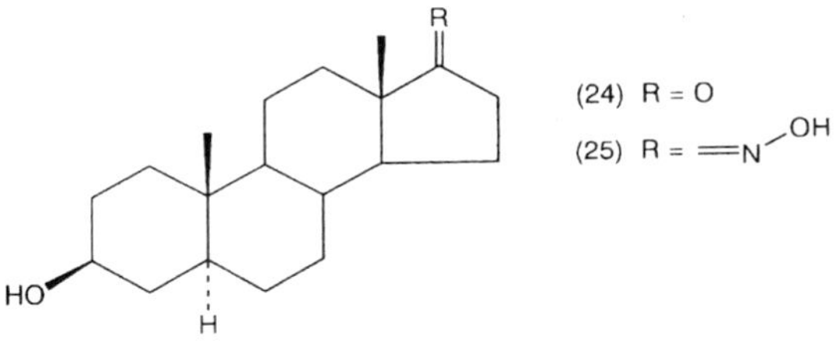

(24) R = O

(25) R = =N$^{\diagup OH}$

ing pyridine (3 cm^3) was heated under reflux for 3 h. The mixture was poured onto ice and the white precipitate which formed was filtered and washed with water to afford the oxime (17-hydroxyimino-5α-androstan-3β-ol, **25**) (2.9 g) which crystallised from ethanol as plates, m.p. 184°C–185°C.

8.2.3   The preparation of an enol ether of a β-diketone

The substrate (dihydroresorcinol, **26**) (5 g) and silver nitrate (7.5 g) were dissolved in water (40 cm^3), and 1 M sodium hydroxide was slowly added

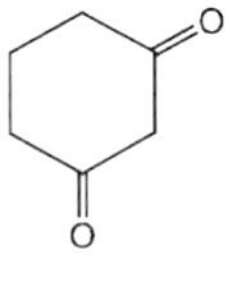

(26)

to adjust the pH to 6. The addition of excess sodium hydroxide should be avoided. The silver salt of dihydroresorcinol was collected by filtration and dried with protection from light. The silver salt was suspended in toluene (50 cm^3) and heated under reflux under a Dean and Starke trap to remove the water. The volume was maintained at 50 cm^3 by the further addition of toluene. Ethyl iodide (8 g) was added and the refluxing continued for 15 min. The solution was cooled, filtered and the product (dihydroresorcinol ethyl ether, **27**) recovered by distillation under high vacuum, boiling point (b.p.) 95°C–100°C at 1 mm Hg pressure.

(27)

8.2.4 The preparation of a pyrrolidine enamine

A mixture of the substrate (cyclohexanone, **28**) (6 g) and pyrrolidine (7 g) in toluene (25 cm^3) were heated in a nitrogen atmosphere under reflux in a

(28)

Dean and Starke apparatus. Heating was continued until the evolution of water ceased (approximately 4 h). The excess pyrrolidine and toluene were evaporated under vacuum and the product (pyrrolidinocyclohex-1-ene, **29**) was distilled under vacuum, b.p. 110°C–112°C at 15 mm Hg pressure.

(29)

8.2.5 The preparation of a hydroxymethylene ether

The substrate (2-methylcyclohexanone, **30**) (11.2 g) was added under nitrogen to a suspension of ice-cold sodium methoxide (from 4.6 g sodium) in dry toluene (75 cm^3). Ethyl formate (14.8 g) was then added and the mixture was allowed to stand at room temperature for 8 h. Ice-water (100 cm^3) was added and the layers were separated. The organic layer was washed with dilute sodium hydroxide. The aqueous layers were combined, washed with diethylether and acidified with hydrochloric acid. The resultant oily suspension was saturated with salt and extracted with diethylether. The extract was dried over sodium sulfate and the solvent evaporated. The product (2-methyl-6-hydroxymethylenecyclohexanone, **31**) was purified by distillation, b.p. 90°C–93°C at 15 mm Hg pressure. The hydroxymethyleneketone (7.5 g) and redistilled isobutanol (6 g) in dry toluene (30 cm^3) containing toluene-*p*-sulfonic acid (100 mg) were heated under reflux in a Dean and Starke apparatus for 4 h. The toluene solution was washed with dilute sodium hydroxide and then dried over sodium sulfate. The solvent was evaporated and the residue was distilled to give the product (6-isobutylenol ether of 2-methylcyclohexanone, **32**), b.p. 108°C–112°C at 0.5 mm Hg pressure.

(30)   (31)   (32)

8.3 The protection of carboxylic acids

The chemistry of methods for the protection of carboxylic acids is described in chapter 4. The commonest method utilises the methyl ester. Methyl esters can be prepared under acid-catalysed conditions with hydrochloric acid as a catalyst. It is sometimes necessary to generate a small amount of hydrogen chloride in methanol for this purpose. Rather than add hydrochloric acid, a controlled amount can be generated by adding a measured amount of acetyl chloride to methanol. An alternative method is to mix a weighed amount of hydroxylamine hydrochloride with a small amount of acetone in methanol. Gibberellic acid (**33**) is a

(33) R = H

(34) R = Me

relatively unstable acid which undergoes rearrangement under acidic and basic conditions. A number of relatively mild conditions have been used to prepare the methyl ester (34) and these are given in the following procedures as examples of routine methylations.

8.3.1 Methylation with diazomethane

Diazomethane must be prepared in an efficient fume cupboard behind a safety screen in apparatus that does not contain ground-glass joints or sharp surfaces. Since it is both toxic and explosive it should not be prepared in large quantities. Excess material should be destroyed by the careful addition of acetic acid. Although a solution of diazomethane in ether may be kept in a refrigerator for a limited period of time, its toxicity and stability should be borne in mind. Furthermore, diazomethane decomposes to produce a polymer which can make crystallisations difficult.

A solution of potassium hydroxide (5 g) in water (10 cm^3) and ethanol (25 cm^3) was placed in a round-bottomed distillation flask fitted with a dropping funnel and a condenser. Two receivers (the second as a trap) each containing ether (25 cm^3) were connected in series to the condenser and cooled in ice. The distillation flask was heated to 60°C on a water bath. A solution of Diazald® (toluene-p-sulfonyl-N-methyl-N-nitroso-amide) (21.4 g) in ether (130 cm^3) was added slowly over a period of 30 min through the dropping funnel to the potassium hydroxide solution so that the yellow solution of ethereal diazomethane gently distilled out and was collected in the first receiver. When the addition was complete a further amount of ether (20 cm^3) was added and the distillation was continued until the distillate was colourless.

The substrate, gibberellic acid (33) (1 g), dissolved in methanol (25 cm^3) in a conical flask in a fume cupboard, was cooled in ice. It was treated with ethereal diazomethane until the yellow colour persisted. The solution was left to stand for a few minutes and then a drop of acetic acid was added to destroy the excess reagent. The solvent was evaporated to give the product, methyl gibberellate (34) (1 g), which was recrystallised from ethyl acetate:light petroleum as needles, m.p. 207°C–209°C.

8.3.2 Methylation with dimethylformamide dimethylacetal

The substrate, gibberellic acid (33) (500 mg), was dissolved in dry tetrahydrofuran (20 cm^3) and treated with dimethylformamide dimethylacetal (15 cm^3) under reflux for 2 h. The solvents were removed *in vacuo* to give a gum which was chromatographed on silica. Elution with 5% methanol in ethyl acetate gave the 3-formyl-7-methyl ester of gibberellic acid (300 mg) which was recrystallised from acetone as needles, m.p. 168°C–170°C.

8.3.3   Methylation with caesium fluoride and methyl iodide

A mixture of the substrate, gibberellic acid (**33**) (2 g), dry caesium fluoride (2 g) and methyl iodide (2 g) in dry dimethylformamide (15 cm^3) was stirred at room temperature for 24 h. The mixture was poured into aqueous sodium hydrogen carbonate and the product recovered in ethyl acetate. The extract was dried over sodium sulfate and the solvent was evaporated to give the product, methyl gibberellate (**34**) (1.7 g), which was recrystallised from ethyl acetate:light petroleum as needles, m.p. 207°C–209°C.

8.4   The protection of the amino group

The chemistry underlying the protection of the amino group has been described in chapter 5. The protection of aromatic amines as their acetyl derivatives, or as part of a pyrrole ring, involves relatively general conditions. However, the protection of the amino group in amino acids has been the subject of considerable effort to find the mildest conditions for specific amino acids which do not perturb the asymmetric centre of the amino acid. Although methods are given for the protection of the secondary amine of proline, specialist texts should be consulted for other amino acids.

8.4.1   The preparation of acetanilide

Aniline (**35**) (5 g) was suspended in water (25 cm^3), and acetic anhydride (5 cm^3) was added. The mixture became hot and the temperature was maintained at 80°C for 20 min before being allowed to cool. The product (**37**) was collected, m.p. 114°C.

(35) R = H   (37) R = H
(36) R = CO$_2$H   (38) R = CO$_2$H

For aromatic amines that are poorly soluble in water, acetone may be used as a co-solvent.

8.4.2 Acetylation of anthranilic acid

Anthranilic acid (36) (1.4 g) was dissolved in a solution of sodium hydroxide (3 g) in water (100 cm^3), and acetic anhydride (5 cm^3) was added. The mixture was stirred for 1 h and then cautiously acidified with dilute hydrochloric acid to yield o-acetamidobenzoic acid (38) (1 g) as prisms, m.p. 185°C–186°C.

8.4.3 The protection of p-bromoaniline as a dimethylpyrrole

A mixture of p-bromoaniline (39) (2.8 g) and hexane-2,5-dione (2 g) in toluene (20 cm^3) containing glacial acetic acid (0.5 cm^3) was heated under reflux using a Dean and Starke trap for 3 h. The mixture was cooled, diluted with ether, washed with dilute hydrochloric acid, aqueous sodium hydrogen carbonate, brine and dried over sodium sulfate. The solvent was evaporated to give 1-(4-bromophenyl)-2,5-dimethylpyrrole (40) (3.5 g), m.p. 74°C–75°C. Deprotection was achieved by heating the pyrrole with excess hydroxylamine hydrochloride in aqueous ethanol for 24 h.

(39)   (40)

8.4.4 The preparation of a carbobenzyloxy derivative

The substrate (L-proline) (41) (4.6 g) was dissolved in 2M sodium hydroxide solution (20 cm^3) and the solution was cooled in an ice-bath. Benzyl chloroformate (7 g) and 2M sodium hydroxide solution (20 cm^3) were added alternately and in portions to the stirred solution over a period of 30 minutes. The mixture was then stirred at room temperature for an hour. It was extracted with ether to remove the excess reagents and then acidified to pH 3 with dilute hydrochloric acid and again extracted with ether. The second extract was washed with water and dried over sodium sulfate. The solvent was evaporated to give the product (carbo-benzyloxy-L-proline) (42) (6 g) with a melting point of 75°C.

8.4.5   The preparation of a t-butoxycarbonyl derivative

The substrate (L-proline, **41**) (1.15 g) was dissolved in dry dimethylsulf-oxide (10 cm³) and treated with 1,1,3,3-tetramethylguanidine (1.15 g) and

(41) R = H

(42) R = Ph

(43) R =

t-butylphenylcarbonate (2.14 g). The mixture was stirred for 3 h and poured into water (100 cm³). The aqueous phase was extracted with ether and then carefully acidified to pH 3 with dilute hydrochloric acid. The product was recovered in ethyl acetate. The extracts were washed with water and dried over sodium sulfate. The solvent was evaporated *in vacuo* to give the derivative (t-butyloxycarbonyl-L-proline, **43**) (1.5 g) as needles, m.p. 130°C.

8.4.6   The preparation of an N-phthalyl derivative

The substrate (L-phenylalanine, **44**) (1.6 g) and powdered phthalic anhydride (1.5 g) were heated in toluene (20 cm³) containing triethylamine (1 cm³) under azeotropic distillation in a Dean and Starke apparatus for 2 h. Fresh toluene was added to replace the toluene that was distilled. The solvent was then removed *in vacuo* and the residue was powdered under water containing dilute hydrochloric acid (50 cm³). The residue was fil-tered and washed with water to give the product (N-phthalyl-L-phenyl-alanine, **45**) (2 g), m.p. 178°C.

(44)

(45)

Appendix

Some common abbreviations for protecting groups and reagents

Ac	Acetyl
Alloc	Allyloxycarbonyl
Ar	Aryl
9-BBN	9-Borabicyclo[3,3,1]nonane
BIPSOP	Bis[(triisopropylsilyl)oxy]pyrrole
Bn	Benzyl
Boc	t-Butyloxycarbonyl (sometimes t-BOC)
BOM	Benzyloxymethyl
Bpoc	Diphenyloxycarbonyl
Brs	4-Bromobenzenesulfonate
Bt	1-Benzotriazoyl
Bu	Butyl (t-Bu or ^{t}Bu is t-butyl)
Bz	Benzoyl
CAN	Ceric ammonium nitrate
Cbz	Benzyloxycarbonyl (carbobenzyloxy)
CSA	Camphorsulfonic acid
Cys	Cysteine
DABCO	1,4-Diazabicyclo[2,2,2]octane
DAST	Diethylaminosulfur trifluoride
DBN	1,5-Diazabicyclo[4,3,0]non-5-ene
DBU	1,8-Diazabicyclo[5,4,0]undec-7-ene
DCC	Dicyclohexylcarbodiimide
DDQ	2,3-Dichloro-5,6-dicyano-1,4-benzoquinone
DEAD	Diethyl azodicarboxylate
DIBAL	Diisobutylaluminium hydride
DMAP	4-N,N-Dimethylaminopyridine
DMB	3,4-Dimethoxybenzyl
DME	Dimethoxyethane
DMF	N,N-Dimethylformamide
DMFDMA	N,N-Dimethylformamide dimethylacetal
DMPU	1,3-Dimethyl-3,4,5,6-tetrahydro-2(1H)-pyrimidinone
DMS	Dimethylsulfide
DMSO	Dimethylsulfoxide
DNB	Dinitrobenzyl (or 4,4′-dinitrobenzhydryl)
DNP	Dinitrophenyl (sometimes dinitrophenylhydrazone)

DPM	Diphenylmethyl
DPMS	Diphenylmethylsilyl
DPTC	O,O'-di(2-pyridyl)thiocarbonate
EDTA	Ethylenediaminetetraacetic acid
EE	2-Ethoxyethyl
Et	Ethyl
Fm	9-Fluorenylmethyl
Fmoc	9-Fluorenylmethoxycarbonyl
GUM	Guaiacylmethoxy
HMDS	1,1,1,3,3,3-Hexamethyldisilazane
HMPA	Hexamethylphosphoramide
HMPT	Hexamethylphosphorustriamide
HOBT	1-Hydroxybenzotriazole
hν	light (often ultraviolet irradiation)
IPDMS	Isopropyldimethylsilyl
LAH	Lithium aluminium hydride
MBS	4-Methoxybenzenesulfonyl
MCPBA	3-Chloroperbenzoic acid
Me	Methyl
MEM	2-Methoxyethoxymethyl
MNNG	1-Methyl-3-nitro-1-nitrosoguanidine
MOM	Methoxymethyl
Ms	Methanesulfonate (mesylate)
MTM	Methylthiomethyl
Mts	Mesitylenesulfonyl
NBS	N-Bromosuccinimide
NPS	2-Nitrophenylsulfenyl
p-AOM	4-Methoxyphenoxymethyl (anisoyloxymethyl)
PCC	Pyridinium chlorochromate
Ph	Phenyl
PMB	4-Methoxybenzyl
POM	4-Pentenyloxymethyl (or pivaloyloxymethyl)
PPTS	Pyridinium toluene-p-sulfonate
Pr	Propyl
Pv	Pivaloate
Pyr	Pyridine
SEM	2-(Trimethylsilyl)ethoxymethyl
Su	Succinimide
TBDMS	t-Butyldimethylsilyl
TBDPS	t-Butyldiphenylsilyl
TEA	Triethylamine
Teoc	2-(Trimethylsilyl)ethoxycarbonyl
TES	Triethylsilyl

TFA	Trifluoroacetyl (or trifluoroacetic acid)
THF	Tetrahydrofuran
THP	Tetrahydropyranyl
TIPS	Triisopropylsilyl
TMEDA	N,N,N′,N′-Tetramethylethylenediamine
TMS	Trimethylsilyl
Tr	Trityl (Triphenylmethyl)
Troc	2,2,2-Trichloroethoxycarbonyl
Ts	Toluene-p-sulfonyl
Z	Benzyloxycarbonyl

References

1. J.F.W. McOmie, *Protective Groups in Organic Chemistry*, Plenum Press, New York, 1973.
2. T.W. Greene and P.G.M. Wuts, *Protective Groups in Organic Synthesis*, 2nd edn., Wiley-Interscience, New York, 1991.
3. P.J. Kocienski, *Protective Groups in Organic Chemistry*, G. Thieme Verlag, Stuttgart, 1994.
4. E.J. Corey, A.K. Long, T.W. Greene and J.W. Miller, 1985, *J. Org. Chem.*, 50, 1920.
5. K. Jarowicki and P. Kocienski, 1995, *Contemporary Organic Synthesis*, 2, 315; 1996, 3, 397; 1997, 4, 454; *J. Chem. Soc., Perkin Trans.*, 1, 1998, 4005; 1999, 1589.
6. M. Schelhaas and H. Waldmann, 1996, *Angew. Chem., Int. Ed. Engl.*, 35, 2056.
7. G. Barany and F. Albericio, 1985, *J. Am. Chem. Soc.*, 107, 4936; G. Barany and R.B. Merrifield, 1977, *J. Am. Chem. Soc.*, 99, 7363.
8. M. Lalonde and T.H. Chan, 1985, *Synthesis*, 817.
9. T.D. Nelson and R.D. Crouch, 1996, *Synthesis*, 1031.
10. J. Muzart, 1993, *Synthesis*, 11.
11. G.A. Olah and S.C. Narang, 1982, *Tetrahedron*, 38, 2225.
12. V.N.R. Pillai, 1980, *Synthesis*, 1.
13. H.J.E. Loewenthal, 1959, *Tetrahedron*, 6, 269.
14. M.V. Bhatt and S.U. Kulkarni, 1983, *Synthesis*, 249.
15. M. Node, K. Ohta, T. Kajimoto, K. Nishida, E. Fujita and K. Fuji, 1983, *Chem. and Pharm. Bull (Jpn)*, 31, 4178.
16. J.C. Irvine and E.L. Hirst, 1923, *J. Chem. Soc.*, 123, 518.
17. H.M.S. Kumar, B.V.S. Reddy, P.K. Mohanty and J.S. Yadav, 1997, *Tetrahedron Letters*, 38, 3619.
18. H.J. Liu, J. Yip and K.S. Shia, 1997, *Tetrahedron Letters*, 38, 2253.
19. T. Iwashige and H. Saeki, 1967, *Chem. and Pharm. Bull. (Jpn)*, 15, 1803.
20. R. Kuhn, I. Low and H. Trischmann, 1957, *Chem. Ber.*, 90, 203.
21. Y. Rabinsohn and H.G. Fletcher, 1967, *J. Org. Chem.*, 32, 3452.
22. M. Smith, D.H. Rammler, I.H. Goldberg and H.G. Khorana, 1962, *J. Am. Chem. Soc.*, 84, 430.
23. T. Onoda, R. Hirai and S. Iwasaki, 1997, *Tetrahedron Letters*, 38, 1443.
24. A.R. Vamo and W.A. Szarek, 1995, *Synlett.*, 1157.
25. G. Just, Z.Y. Wang and L. Chen, 1988, *J. Org. Chem.*, 53, 1030.
26. A. Sikrishna, R. Viswajanani, J.A. Sattigeria and D. Vijaykumar, 1995, *J. Org. Chem.*, 60, 5961.
27. S.M. Kadam, S.K. Nayak and A. Banerji, 1992, *Tetrahedron Letters*, 33, 5129.
28. J.S. Yadav, S. Chandrasekhar, G. Sumithra and R. Kache, 1996, *Tetrahedron Letters*, 37, 6603.
29. H.S. Cho, J. Yu and J.R. Falck, 1994, *J. Am. Chem. Soc.*, 116, 8354.
30. W.E. Parham and E.L. Anderson, 1948, *J. Am. Chem. Soc.*, 70, 4187.
31. A.C. Ott, M.F. Murray and R.L. Pederson, 1952, *J. Am. Chem. Soc.*, 74, 1239.
32. H.B. Henbest, E.R.H. Jones and I.M.S. Walls, 1950, *J. Chem. Soc.*, 3646.
33. K.R. Kloetstra and H. Van Bekkum, 1995, *J. Chem. Research (S)*, 26.
34. Z.H. Zhang, T.S. Li, T.S. Jin and J.X. Wang, 1998, *J. Chem. Research (S)*, 152.
35. I. Mohammadpoor-Baltork and B. Kharamesh, 1998, *J. Chem. Research (S)*, 146.
36. T. Schlama, V. Gouverneur and C. Mioskowski, 1997, *Tetrahedron Letters*, 38, 3517.
37. S. Ma and L.M. Venanzi, 1993, *Tetrahedron Letters*, 34, 5270; 1993, 34, 8071.
38. K. Fuji, S. Nakano and E. Fujita, 1975, *Synthesis*, 276.
39. G.A. Olah, A. Husain and S.C. Narang, 1983, *Synthesis*, 896.

40. E.J. Corey and M.G. Bock, 1975, *Tetrahedron Letters*, 3269.
41. E.J. Corey, J.L. Gras and P. Ulrich, 1976, *Tetrahedron Letters*, 809.
42. J.H. Rigby and J.Z. Wilson, 1984, *Tetrahedron Letters*, 25, 1429.
43. B. Loubinoux, G. Coudert and G. Guillaumet, 1981, *Tetrahedron Letters*, 22, 1973.
44. Z. Wu, D.R. Mootoo and B. Fraser-Reid, 1988, *Tetrahedron Letters*, 29, 4549.
45. Y. Masaki, I. Iwata, I. Mukai, H. Oda and H. Nagashima, 1989, *Chemistry Letters*, 659.
46. B.H. Lipshutz and J.J. Pegram, 1980, *Tetrahedron Letters*, 21, 3343.
47. R.B. Woodward, K. Heusler, J. Gosteli, P. Naegeli, W. Oppolzer, R. Ramage, S. Ranganathan and H. Vorbruggen, 1966, *J. Am. Chem. Soc.*, 88, 852.
48. For a review, see M. Lalonde and T.H. Chan, 1985, *Synthesis*, 817; T.D. Nelson and R.D. Crouch, 1996, *Synthesis*, 1031.
49. E.J. Corey and K.Y. Yi, 1992, *Tetrahedron Letters*, 33, 2289.
50. A.S.Y. Lee, H.C. Yeh and J.J. Shie, 1998, *Tetrahedron Letters*, 39, 5249.
51. T.W. Hart, D.A. Metcalfe and F. Scheinmann, 1979, *J. Chem. Soc., Chem. Commun.*, 156.
52. E.J. Corey and A. Venkateswarlu, 1972, *J. Am. Chem. Soc.*, 94, 6190.
53. S. Hanessian and P. Lavallee, 1975, *Canad. J. Chem.*, 53, 2975.
54. E.J. Corey and S. Choi, 1993, *Tetrahedron Letters*, 34, 6969.
55. J.W. Gillard, R. Fortin, H.E. Morton, C. Yoakim, C.A. Quesnelle, S. Daignaault and Y. Guindon, 1988, *J. Org. Chem.*, 53, 2602.
56. T. Schmittberger and D. Uguen, 1995, *Tetrahedron Letters*, 36, 7445.
57. M.A. Brook, C. Gottardo, S. Balduzzi and M. Mohamad, 1997, *Tetrahedron Letters*, 38, 6997.
58. G. Hofle, W. Steglich and H. Vorbruggen, 1978, *Angew. Chem. Int. Ed. Engl.*, 17, 569.
59. K. Ishihara, H. Kurihara and H. Yamamoto, 1993, *J. Org. Chem.*, 58, 3791.
60. K. Ishihara, M. Kubota, H. Kurihara and H. Yamamoto, 1996, *J. Org. Chem.*, 61, 4560.
61. H.X. Li, T.S. Li and T.H. Ding, 1997, *J. Chem. Soc., Chem. Commun.*, 1389.
62. H.J.E. Loewenthal, 1959, *Tetrahedron*, 6, 269.
63. J.R. Hanson, C. Uyanik and K. Yildrim, 1998, *J. Chem. Research (S)*, 580.
64. F. Reber, A. Lardon and T. Reichstein, 1954, *Helv. Chim. Acta*, 37, 45.
65. C.B. Reese and J.C.M. Stewart, 1968, *Tetrahedron Letters*, 4273.
66. J.R. Hanson, P.B. Reese and H.J. Wadsworth, 1984, *J. Chem. Soc., Perkin Trans.*, 1, 2941.
67. F. Santoyo-Gonzalez, F. Garcis-Calvo-Flores, J.I. Garcia, R. Robles-Diaz and A. Vargas-Berenguel, 1994, *Synthesis*, 97.
68. L.F. Fieser, J.E. Herz, M.W. Klohs, M.A. Romero and T. Utne, 1952, *J. Am. Chem. Soc.*, 74, 3309.
69. T.B. Windholz and D.B.R. Johnston, 1967, *Tetrahedron Letters*, 2555.
70. M. Bessodes and C. Boukarim, 1996, *Synlett.*, 1119.
71. R. Barker and D.L. MacDonald, 1960, *J. Am. Chem. Soc.*, 82, 2301.
72. A.J. Showler and P.A. Darley, 1967, *Chemical Reviews*, 67, 427.
73. S.V. Ley, M. Woods and A. Zanotti-Gerosa, 1992, *Synthesis*, 52.
74. S.V. Ley, H.W.M. Priepke and S.L. Warriner, 1994, *Angew. Chem., Int. Ed. Engl.*, 33, 2290.
75. R.M. Burk and M.B. Roof, 1993, *Tetrahedron Letters*, 34, 395.
76. H. Eckert and B. Forster, 1987, *Angew. Chem. Int. Ed. Engl.*, 26, 894.
77. W.G. Overend, M. Stacey and L.F. Wiggins, 1949, *J. Chem. Soc.*, 1358.
78. G.W. Breton, M.J. Kurtz and S.L. Kurtz, 1997, *Tetrahedron Letters*, 38, 3825.
79. F. de Angelis, M. Marzi, P. Minetti, D. Misiti and S. Muck, 1997, *J. Org. Chem.*, 62, 4159.
80. H.W. Lee and Y. Kishi, 1985, *J. Org. Chem.*, 50, 4402.
81. M.F. Mosquera, M. Martin-Lomas and J.L. Chiara, 1998, *Tetrahedron Letters*, 39, 5085.
82. S.W. Garrett, C. Liu, A.M. Riley and B.V.L. Potter, 1998, *J. Chem. Soc., Perkin Trans.*, 1, 1367.
83. R.A.W. Johnstone and M.E. Rose, 1979, *Tetrahedron*, 35, 2169.

84. J.W.F. McOmie, M.L. Watts and D.E. West, 1968, *Tetrahedron*, 24, 2289.
85. F.E. King, T.J. King and L.C. Manning, 1953, *J. Chem. Soc.*, 3932.
86. R.R. Scheline, 1966, *Acta Chem. Scand.*, 20, 1182.
87. W. Bonthrone and J.W. Cornforth, 1969, *J. Chem. Soc.*, 1202.
88. F.A.J. Meskens, 1981, *Synthesis*, 501.
89. E.C. Taylor and C.S. Chiang, 1977, *Synthesis*, 467.
90. L. Goodman and B.R. Baker, 1961, *J. Org. Chem.*, 26, 1156.
91. W.G. Dauben, R.R. Ollman and S.C. Wu, 1994, *Tetrahedron Letters*, 35, 2149.
92. Y. Kamitori, M. Hojo, R. Masuda and T. Yoshida, 1985, *Tetrahedron Letters*, 26, 4767.
93. M. Sulzbacher, E. Bergmann and E.R. Pariser, 1948, *J. Am. Chem. Soc.*, 70, 2827.
94. W.J. Dauben, B. Laken and H.J. Ringold, 1954, *J. Am. Chem. Soc.*, 76, 1359.
95. H. Meerwein, W. Florian and N. Schon, 1961, *Annalen*, 641, 1.
96. O. Isler, H. Lindlar, M. Montaron, R. Ruegg, G. Saucy and P. Zeller, 1956, *Helv. Chim. Acta*, 39, 2041.
97. S.S. Elmorsy, M.V. Bhatt and A. Pelter, 1992, *Tetrahedron Letters*, 33, 1657.
98. A.K. Mandal, P.Y. Shrotri and A.D. Ghogare, 1986, *Synthesis*, 221.
99. B.M. Choudary and P.N. Reddy, 1995, *Synlett.*, 959.
100. S. Bhat, A.R. Ramesha and S. Chandrasekaran, 1995, *Synlett.*, 329.
101. E.J. Corey, E.J. Trybulski and J.W. Suggs, 1976, *Tetrahedron Letters*, 4577.
102. E.J. Corey and R.A. Ruden, 1973, *J. Org. Chem.*, 38, 834.
103. B.M. Lillie and M.A. Avery, 1994, *Tetrahedron Letters*, 35, 969.
104. Y. Oikawa, T. Yoshioka and O. Yonemitsu, 1982, *Tetrahedron Letters*, 23, 889.
105. S. Kim, Y.G. Kim and D. Kim, 1992, *Tetrahedron Letters*, 33, 2565.
106. C. Djerassi and M. Gorman, 1953, *J. Am. Chem. Soc.*, 75, 3704.
107. H. Zinner, 1950, *Chem. Ber.*, 83, 275.
108. J.R. Williams and G.M. Sarkisian, 1974, *Synthesis*, 32.
109. G. Majetich, M. Behnke and K. Hull, 1985, *J. Org. Chem.*, 50, 3615.
110. D. Evans, L.K. Truesdale, K.G. Grimm and S.L. Nesbitt, 1977, *J. Am. Chem. Soc.*, 99, 5009.
111. R.S. Varma and R.K. Saini, 1997, *Tetrahedron Letters*, 38, 2623.
112. C. Djerassi, M. Shamma and T.Y. Khan, 1958, *J. Am. Chem. Soc.*, 80, 4723.
113. E.L. Eliel and S. Krishnamurthy, 1965, *J. Org. Chem.*, 30, 848.
114. M. Prato, U. Quintily, G. Scorrano and A. Sturaro, 1982, *Synthesis*, 679.
115. G.A. Olah, A.K. Mehrotra and S.C. Narang, 1982, *Synthesis*, 151.
116. C.S. Rao, M. Chandrasekharam, H. Ila and H. Junjappa, 1992, *Tetrahedron Letters*, 33, 8163.
117. M. Kamata, H. Otogawa and E. Hasegawa, 1991, *Tetrahedron Letters*, 32, 7421.
118. S.A. Patwardhan and S. Dev, 1974, *Synthesis*, 348.
119. R. Wohl, 1974, *Synthesis*, 38.
120. S. Danishefsky and T. Kitahara, 1974, *J. Am. Chem. Soc.*, 96, 7807.
121. G. Stork and R.L. Danheiser, 1973, *J. Org. Chem.*, 38, 1775.
122. J.L. Johnson, M.E. Herr, J.C. Babcock, H.E. Fonken, J.E. Stafford and F.W. Heyl, 1956, *J. Am. Chem. Soc.*, 78, 430.
123. J.E. McMurry, 1968, *J. Am. Chem. Soc.*, 90, 6821.
124. J.G. Lee, K.H. Kwak and J.P. Hwang, 1990, *Tetrahedron Letters*, 31, 6677.
125. D.W. Chen and Z.C. Chen, 1994, *Synthesis*, 773.
126. J.M. Khurana, H. Ray and P.K. Sahoo, 1994, *Bull. Chem. Soc. (Jpn)*, 67, 1091.
127. R.S. Varma and H.M. Meshram, 1997, *Tetrahedron Letters*, 38, 7973.
128. E.B. Hershberg, 1948, *J. Org. Chem.*, 13, 542.
129. R. Ballini, G. Bosica, R. Maggi and G. Sartori, 1997, *Synlett.*, 795.
130. H. Goda, H. Ihara, C. Hirayama and M. Sato, 1994, *Tetrahedron Letters*, 35, 1565.

131. M. Avaro, J. Levisalles and H. Rudler, 1969, *J. Chem. Soc., Chem. Commun.*, 445.
132. D.H.R. Barton, J.C. Jaszberenyi, W. Liu and T. Shinada, 1996, *Tetrahedron*, 52, 14673.
133. D.A. Evans, L.K. Truesdale and G.L. Carroll, 1973, *J. Chem. Soc., Chem. Commun.*, 55.
134. D.A. Evans, J.M. Hoffman and L.K. Truesdale, 1973, *J. Am. Chem. Soc.*, 95, 5822.
135. F.E. King, T.J. King and J.G. Topliss, 1957, *J. Chem. Soc.*, 919.
136. E. Haslam, 1980, *Tetrahedron*, 36, 2409.
137. J.I. Trujillo and A.S. Gopalan, 1993, *Tetrahedron Letters*, 34, 7355.
138. R.F. Abdulla and R.S. Brinkmeyer, 1979, *Tetrahedron*, 35, 1675.
139. T.J. De Boer and H.J. Backer, 1954, *Rec. Trav. Chim.*, 73, 229.
140. O. Mitsunobu, 1981, *Synthesis*, 1.
141. T. Sato, J. Otera and H. Nozaki, 1992, *J. Org. Chem.*, 57, 2166.
142. G. Kokotos and A. Chiou, 1997, *Synthesis*, 168.
143. M. Faure-Tromeur and S.J. Zard, 1998, *Tetrahedron Letters*, 39, 7301.
144. J. Inanaga, K. Hirata, H. Saeki, T. Katsuki and M. Yamaguchi, 1979, *Bull. Chem. Soc. (Jpn)*, 52, 1989.
145. B. Neises and W. Steglich, 1978, *Angew. Chem. Int. Ed., Engl.*, 17, 522.
146. J.C. Sheehan and J.J. Hlavka, 1956, *J. Org. Chem.*, 21, 439.
147. M.K. Dhaun, R.K. Olsen and K. Ramasamy, 1982, *J. Org. Chem.*, 47, 1962.
148. T. Mukaiyama, 1979, *Angew. Chem. Int. Ed. Engl.*, 18, 707.
149. M. Saitoh, I. Shiina and T. Mukaiyama, 1998, *Chemistry Letters*, 479.
150. J.J. Folmer and S.M. Weinreb, 1993, *Tetrahedron Letters*, 34, 2737.
151. M.A. Brook and T.H. Chan, 1983, *Synthesis*, 201.
152. P.D. Raddo, 1986, *J. Chem. Educ.*, 1034.
153. E.F.V. Scriven, 1983, *Chem. Soc. Reviews*, 12, 129.
154. C. Salomon, E.G. Mata and O.A. Mescaretti, 1993, *Tetrahedron*, 49, 3691.
155. P.A. Bartlett and W.S. Johnson, 1974, *Tetrahedron Letters*, 39, 1427.
156. M. Node, K. Nishida, M. Sai and E. Fujita, 1978, *Tetrahedron Letters*, 5211.
157. G.A. Olah, S.C. Narang, G.F. Salem and B.G. Gupta, 1981, *Synthesis*, 142.
158. A.M. Felix, 1974, *J. Org. Chem.*, 39, 1427.
159. M. Node, K. Nishida, M. Sai, K. Fuji and E. Fujita, 1981, *J. Org. Chem.*, 46, 1991.
160. M.S. Bernatowicz, H.G. Chao and G.S. Matsueda, 1994, *Tetrahedron Letters*, 35, 1651.
161. R. Paredes, F. Agudelo and G. Taborda, 1996, *Tetrahedron Letters*, 37, 1965.
162. M.S. Anson and J.G. Montana, 1994, *Synlett.*, 219.
163. S.W. Wright, D.L. Hagemann, A.S. Wright and L.D. McClure, 1997, *Tetrahedron Letters*, 38, 7345.
164. W.C. Chen, M.D. Vera and M.M. Joullie, 1997, *Tetrahedron Letters*, 38, 4025.
165. Y. Kita, H. Maeda, F. Takahashi, S. Fukui, T. Ogawa and K. Hatayama, 1994, *Chem. Pharm. Bull. (Jpn)*, 42, 147.
166. A.D. Proud, J.C. Prodger and S.L. Flitsch, 1997, *Tetrahedron Letters*, 38, 7243.
167. J.C. Roberts, H. Gao, A. Gopalsamy, A. Kongsjahju and R.J. Patch, 1997, *Tetrahedron Letters*, 38, 355.
168. E.P. Serebryakov, L.M. Suslova and V.F. Kucherov, 1978, *Tetrahedron*, 34, 345.
169. B. Kundu, 1992, *Tetrahedron Letters*, 33, 3193.
170. T. Mandai, M. Imaji, H. Takada, M. Kawata, J. Nokami and J. Tsuji, 1989, *J. Org. Chem.*, 54, 5395.
171. T. Murayama, A. Yoshida, T. Kobayashi and T. Miura, 1994, *Tetrahedron Letters*, 35, 2271.
172. J. Cossy, A. Albouy, M. Scheloske and D.G. Pardo, 1994, *Tetrahedron Letters*, 35, 1539.
173. G. Voss and H. Gerlach, 1983, *Helv. Chim. Acta*, 66, 2294.
174. B. Rousseau, J.P. Beaucourt and L. Pichat, 1982, *Tetrahedron Letters*, 23, 2183.
175. T.P. Mawhinney and M.A. Madson, 1982, *J. Org. Chem.*, 47, 3336.

176. S. Djuric, J. Venit and P. Magnus, 1981, *Tetrahedron Letters*, 22, 1787.
177. T. Guggenheim, 1984, *Tetrahedron Letters*, 25, 1253.
178. R.P. Bonar-Law, A.P. Davis and B.J. Dorgan, 1990, *Tetrahedron Letters*, 31, 6721.
179. E.C. Taylor, O. Vogl and P.K. Loeffler, 1959, *J. Am. Chem. Soc.*, 81, 2479.
180. S.P. Bruekelman, S. Leach, G.D. Meakins and M.D. Tirel, 1984, *J. Chem. Soc., Perkin Trans.*, 1, 2801.
181. J.A. Ragan, T.W. Makowski, M.J. Castaldi and P.D. Hill, 1998, *Synthesis*, 1599.
182. S.F. Martin and C. Limberakis, 1997, *Tetrahedron Letters*, 38, 2617.
183. F. Koniuszy, P.F. Wiley and K. Folkers, 1949, *J. Am. Chem. Soc.*, 71, 875.
184. J.C. Sheehan and D.D.H. Yang, 1958, *J. Am. Chem. Soc.*, 80, 1154.
185. S. Vincent, S. Mons, L. Lebeau and C. Mioskowski, 1997, *Tetrahedron Letters*, 38, 7527.
186. A.E. Smith and K.J.P. Orton, 1908, *J. Chem. Soc.*, 93, 1242.
187. S. Hanessian, 1967, *Tetrahedron Letters*, 1549.
188. H. Waldmann, A. Heuser and A. Reidel, 1994, *Synlett.*, 65.
189. H. Waldmann, A. Heuser and S. Schulz, 1996, *Tetrahedron Letters*, 37, 8725.
190. A. Taurog, S. Abraham and I.L. Chaikoff, 1953, *J. Am. Chem. Soc.*, 75, 3473.
191. D. Xu, K. Prasad, O. Repic and T.J. Blacklock, 1995, *Tetrahedron Letters*, 36, 7357.
192. L.A. Carpino and F. Nowshad, 1993, *Tetrahedron Letters*, 34, 7009.
193. G.H.L. Nefkens, G.I. Tesser and R.J.F. Nivard, 1960, *Recl. Trav. Chim. Pays-Bas*, 79, 689.
194. H.R. Ing and R.F.H. Manske, 1926, *J. Chem. Soc.*, 2348.
195. H. Tsubouchi, K. Tsuji and H. Ishikawa, 1994, *Synlett.*, 63.
196. A.G. Schultz, P.J. McCloskey and J.J. Court, 1987, *J. Am. Chem. Soc.*, 109, 6943.
197. E. Vedejs and S. Lin, 1994, *J. Org. Chem.*, 59, 1602.
198. C.G. Dubois, A. Guggisberg and M. Hesse, 1995, *J. Org. Chem.*, 60, 5969.
199. T. Fujii and S. Sakakibara, 1974, *Bull. Chem. Soc. (Jpn)*, 47, 3146.
200. L. Zervas, D. Borovas and E. Gazis, 1963, *J. Am. Chem. Soc.*, 85, 3660.
201. J. Jones, *Amino Acid and Peptide Synthesis*, Oxford University Press, Oxford, 1992.
202. M. Bergmann and L. Zervas, 1932, *Ber. Dtsch. Chem. Ges.*, 65, 1192.
203. R. Geiger and W. Konig, *The Peptides* Eds. E. Gross and J. Meinhofer, Academic Press, New York, 1981, vol. 3, p. 15.
204. F.H. Carpenter and D.T. Gish, 1952, *J. Am. Chem. Soc.*, 74, 3818.
205. F. Weygand and K. Hunger, 1962, *Chem. Ber.*, 95, 1.
206. J.E. Shields and F.H. Carpenter, 1961, *J. Am. Chem. Soc.*, 83, 3066.
207. F.C. McKay and N.F. Albertson, 1957, *J. Am. Chem. Soc.*, 79, 4686.
208. G.W. Anderson and A.C. McGregor, 1957, *J. Am. Chem. Soc.*, 79, 6180.
209. U. Ragnarsson, S.M. Karlsson and B.E. Sandberg, 1972, *Acta Chem. Scand.*, 26, 2550.
210. M. Itoh, D. Hagiwara and T. Kamiya, 1975, *Tetrahedron Letters*, 4393.
211. P. Sieber and B. Iselin, 1968, *Helv. Chim. Acta*, 51, 614.
212. L.A. Carpino, 1987, *Acc. Chem. Research*, 20, 401.
213. A.R. Brown, S.L. Irving and R. Ramage, 1993, *Tetrahedron Letters*, 34, 7129.
214. G.G.J. Verhart and G.I. Tesser, 1988, *Recl. Trav. Chim. Pays-Bas*, 107, 621.
215. E. Wunsch and R. Spangenberg, 1971, *Chem. Ber.*, 104, 2427.
216. L.A. Carpino, J.H. Tsao, H. Ringsdorf, E. Fell and G. Hettrich, 1978, *J. Chem. Soc., Chem. Commun.*, 358.
217. J. Grimshaw, 1965, *J. Chem. Soc.*, 7136.
218. J.A. Stafford, M.F. Brackeen, D.S. Karanewsky and N.L. Valvano, 1993, *Tetrahedron Letters*, 34, 7873.
219. R. Ledger and F.H.C. Stewart, 1965, *Aust. J. Chem.*, 18, 933.
220. K. Hofmann, W.D. Peckham and A. Rheiner, 1956, *J. Am. Chem. Soc.*, 78, 238.
221. P.M. Scopes, K.B. Walshaw, M. Welford and G.T. Young, 1965, *J. Chem. Soc.*, 782.
222. R.H. Sifferd and V. du Vigneaud, 1935, *J. Biol. Chem.*, 108, 753.

223. J.E.T. Corrie, J.R. Hlubucek and G. Lowe, 1977, *J. Chem. Soc., Perkin Trans.*, 1, 1421.
224. M. Frankel, D. Gertner, H. Jacobsen and A. Zilkha, 1960, *J. Chem. Soc.*, 1390.
225. S. Sakakibara, Y. Shimonishi, Y. Kishida, M. Okada and H. Sugihara, 1967, *Bull. Chem. Soc. (Jpn)*, 40, 2164.
226. S. Akabori, S. Sakakibara, Y. Shimonishi and Y. Nobuhara, 1964, *Bull. Chem. Soc. (Jpn)*, 37, 433.
227. R.G. Hiskey, T. Mizoguchi and H. Igeta, 1966, *J. Org. Chem.*, 31, 1188.
228. F. Dardoize, M. Gaudemar and N. Goasdove, 1977, *Synthesis*, 567.
229. P.J.E. Brownlee, M.E. Cox, B.O. Handford, J.C. Marsden and G.T. Young, 1964, *J. Chem. Soc.*, 3832.
230. R. Camble, R. Purkayasatha and G.T. Young, 1968, *J. Chem. Soc. (C)*, 1219.
231. D.F. Veber, J.D. Milkowski, S.L. Varga, R.G. Denkewalter and R. Hirschmann, 1972, *J. Am. Chem. Soc.*, 94, 5456.
232. M. Ruiz-Gaypo, F. Albericio, E. Pedroso and E. Giralt, 1986, *J. Chem. Soc., Chem. Commun.*, 1501.
233. Y. Ohtsuka, S. Niitsuma, H. Tadokoro, T. Hayashi and T. Oishi, 1984, *J. Org. Chem.*, 49, 2326.
234. A. Berger, J. Noguchi and E. Katchalski, 1956, *J. Am. Chem. Soc.*, 78, 4483.
235. M. Muraki and T. Mizoguchi, 1971, *Chem. Pharm. Bull. (Jpn)*, 19, 1708.
236. A. Fontana, 1975, *J. Chem. Soc., Chem. Commun.*, 976.
237. J.J. Pastuszak and A. Chimiak, 1981, *J. Org. Chem.*, 46, 1868.
238. L.F. Fieser, 1954, *J. Chem. Educ.*, 31, 291.
239. L. Ruzicka and A. Wettstein, 1935, *Helv. Chim. Acta*, 18, 986.
240. A.F. St. Andre, H.B. MacPhillamy, J.A. Nelson, A.C. Shabica and C.R. Scholz, 1952, *J. Am. Chem. Soc.*, 74, 5506.
241. G.A. Olah and G.K.S. Prakash, 1976, *Synthesis*, 607.
242. P.E. Sonnet, 1980, *J. Org. Chem.*, 45, 154.
243. J. MacMillan and C.L. Willis, 1986, *J. Chem. Soc., Perkin Trans.*, 1, 309.
244. J.W. Cornforth, R.H. Cornforth and K.K. Mathew, 1959, *J. Chem. Soc.*, 112.
245. R.N. Barnah, R.P. Sharma and J.N. Barnah, 1983, *Chem. and Ind.*, 524.
246. E.J. Corey and R.A.E. Winter, 1963, *J. Am. Chem. Soc.*, 85, 2677.
247. E.J. Corey, F.A. Carey and R.A.E. Winter, 1965, *J. Am. Chem. Soc.*, 87, 934.
248. S. Kim and K.Y. Yi, 1986, *J. Org. Chem.*, 51, 2613.
249. F.W. Eastwood, K.J. Harrington, J.S. Josan and J.L. Pura, 1970, *Tetrahedron Letters*, 5223.
250. A. Ichihara, K. Oda, M. Kobayashi and S. Sakamura, 1980, *Tetrahedron*, 36, 183.
251. D.H.R. Barton, T. Shioiri and D.A. Widdowson, 1971, *J. Chem. Soc. (C)*, 1968.
252. D.H.R. Barton, A.A.I. Gunatilaka, T. Nakanishi, H. Patin, D.A. Widdowson and B.R. Worth, 1976, *J. Chem. Soc., Perkin Trans.*, 1, 821.
253. L. Van Maele, P.J. De Clercq and M. Vandewalle, 1985, *Tetrahedron*, 41, 141.
254. S.F. Martin, S.R. Desai, G.W. Phillips and A.C. Miller, 1980, *J. Am. Chem. Soc.*, 102, 3294.
255. K.M. Nicholas and R. Pettit, 1971, *Tetrahedron Letters*, 3475.
256. D. Seyferth, M.O. Nestle and A.J. Wehman, 1975, *J. Am. Chem. Soc.*, 97, 7417.
257. T.H. Chan and I. Fleming, 1979, *Synthesis*, 761.
258. T. Floor and P.E. Paterson, 1980, *J. Org. Chem.*, 45, 5006.
259. E. Sakellarios, 1925, *Ber. Dtsch. Chem. Ges.*, 58, 2286.
260. L.H. Welsh, 1941, *J. Am. Chem. Soc.*, 63, 3276.
261. I.M. Hunsberger, D. Lednicer, H.S. Gutowsky, D.L. Bunker and P. Taussig, 1955, *J. Am. Chem. Soc.*, 77, 2466.
262. L.A. Bigelow, 1919, *J. Am. Chem. Soc.*, 41, 1559.

263. T. Kametani and M. Ihara, 1967, *J. Chem. Soc. (C)*, 530.
264. G. Fukata, Y. Kubota, S. Mataka, T. Thiemann and M. Tashiro, 1994, *Bull. Chem. Soc. (Jpn)*, 67, 592.
265. D.D. Pern, W.L.F. Armarego and D.R. Perrin, *Purification of Laboratory Chemicals*, Pergamon, Oxford, 1980.
266. B.S. Furniss, A.J. Hannaford, P.W.G. Smith and A.R. Tatchell, *Vogel's Textbook of Practical Organic Chemistry*, 5th edn, Longman *Scientific and Technical*, Harlow, Essex, 1989.
267. *Organic Syntheses. Collective Volumes 1-8*, John Wiley and Sons, New York, 1941-1993.

Index